THE WOHASCUM COUNTY
PROBLEM BOOK

GEORGE T. GILBERT
MARK I. KRUSEMEYER
LOREN C. LARSON

THE
DOLCIANI MATHEMATICAL EXPOSITIONS

Published by
THE MATHEMATICAL ASSOCIATION OF AMERICA

———

THE WOHASCUM COUNTY PROBLEM BOOK

Gilbert, Krusemeyer, Larson

Published and Distributed by
THE MATHEMATICAL ASSOCIATION OF AMERICA

To Fred and Barbara Gilbert (GG)
To Alice and Elmer Larson (LL)

The DOLCIANI MATHEMATICAL EXPOSITIONS series of the Mathematical Association of America was established through a generous gift to the Association from Mary P. Dolciani, Professor of Mathematics at Hunter College of the City University of New York. In making the gift, Professor Dolciani, herself an exceptionally talented and successful expositor of mathematics, had the purpose of furthering the ideal of excellence in mathematical exposition.

The Association, for its part, was delighted to accept the gracious gesture initiating the revolving fund for this series from one who has served the Association with distinction, both as a member of the Committee on Publications and as a member of the Board of Governors. It was with genuine pleasure that the Board chose to name the series in her honor.

The books in the series are selected for their lucid expository style and stimulating mathematical content. Typically, they contain an ample supply of exercises, many with accompanying solutions. They are intended to be sufficiently elementary for the undergraduate and even the mathematically inclined high-school student to understand and enjoy, but also to be interesting and sometimes challenging to the more advanced mathematician.

DOLCIANI MATHEMATICAL EXPOSITIONS

1. *Mathematical Gems,* Ross Honsberger
2. *Mathematical Gems II,* Ross Honsberger
3. *Mathematical Morsels,* Ross Honsberger
4. *Mathematical Plums,* Ross Honsberger (ed.)
5. *Great Moments in Mathematics (Before 1650)*, Howard Eves
6. *Maxima and Minima without Calculus,* Ivan Niven
7. *Great Moments in Mathematics (After 1650)*, Howard Eves
8. *Map Coloring, Polyhedra, and the Four-Color Problem,* David Barnette
9. *Mathematical Gems III,* Ross Honsberger
10. *More Mathematical Morsels,* Ross Honsberger
11. *Old and New Unsolved Problems in Plane Geometry and Number Theory,* Victor Klee and Stan Wagon
12. *Problems for Mathematicians, Young and Old,* Paul R. Halmos
13. *Excursions in Calculus: An Interplay of the Continuous and the Discrete,* Robert M. Young
14. *The Wohascum County Problem Book,* George T. Gilbert, Mark I. Krusemeyer, Loren C. Larson

PREFACE

Many of the 130 problems in this book were first posed as weekly challenges for undergraduates at either Carleton College or St. Olaf College. All problems presented here are due to the authors; as far as we know, they do not occur in the literature, although they certainly include some variations on classic themes. Some knowledge of linear or abstract algebra is needed for a few of the problems, but most require nothing beyond calculus, and many should be accessible to high school students. However, there is a wide range of difficulty, and some problems require considerable mathematical maturity. For most students, few, if any, of the problems will be routine. We have tried to put easier problems before harder ones, and we expect that nearly everyone will find Problem 85 more difficult than Problem 25. Nevertheless, a particular solver might find 85 easier than 75 or even 65.

Not surprisingly, most of the book is taken up by solutions rather than by problems. When possible, we have tried to include solutions which are elegant and/or instructive in addition to being clear; often, several solutions to the same problem are presented. On the other hand, solvers should be aware that some problems seem to have only messy solutions, and that solutions, especially for the more difficult problems, may run to several pages. Sometimes solutions are preceded by "Ideas," which can serve as motivation or as hints, or followed by "Comments," which often put solutions in a broader perspective. The comments to Problems 38, 94, and 103 indicate related problems which we have so far been unable to solve.

The appendices at the back of the book may be especially helpful to problem solving classes and to teams or individuals preparing for contests such as the Putnam. The first appendix lists the prerequisites for each problem (which may provide hints for possible solution methods!), while the second appendix has the problems arranged by general topic.

So where is Wohascum County, anyway? DeLorme's Minnesota Atlas and Gazetteer™ (DeLorme Mapping Company, Freeport, ME, 1990) does not have a list of counties, and the index of towns does not list Wohascum Center or Lake Wohascum. The closest index entry, for Lake Wobegon, refers the reader to page 97; however, the last map in the atlas is on page 95. Thus we are unable to give you a precise geographic location. Wohascum (pronounced Wo HAS cum) Center is, nonetheless, a thriving, if slightly eccentric, community, and we are grateful to John Johnson of Teapot Graphics for the superb illustrations with which he has brought it to life.

Our heartfelt thanks go, also, to the many people—our parents, spouses, colleagues and friends—whose vital encouragement and support we have been privileged to receive. The following reviewers each read substantial portions of the manuscript, and their thoughtful suggestions led to many improvements: Professors Joe Buhler, Paul Campbell, William Firey, Paul Fjelstad, Steven Galovich, Gerald Heuer, Abraham Hillman, Meyer Jerison, Elgin Johnston, Eugene Luks, Murray Klamkin, Bruce Reznick, the late Ian Richards, Allen Schwenk, John Shue, and William Waterhouse. Beverly Ruedi of the MAA brought unfailing good humor and patience to the nitty-gritty of manuscript production. Finally, our thanks to all past, present, and future students who respond to these and other challenge problems. Enjoy!

George Gilbert
Mark Krusemeyer
Loren Larson

Fort Worth, Texas
Northfield, Minnesota
September 15, 1992

CONTENTS

---

* The page number of the solution appears at the end of each problem.

THE PROBLEMS

1. Find all solutions in integers of $x^3 + 2y^3 = 4z^3$. (p. 33)

2. The Wohascum County Board of Commissioners, which has 20 members, recently had to elect a President. There were three candidates (A, B, and C); on each ballot the three candidates were to be listed in order of preference, with no abstentions. It was found that 11 members, a majority, preferred A over B (thus the other 9 preferred B over A). Similarly, it was found that 12 members preferred C over A. Given these results, it was suggested that B should withdraw, to enable a runoff election between A and C. However, B protested, and it was then found that 14 members preferred B over C! The Board has not yet recovered from the resulting confusion. Given that every possible order of A, B, C appeared on at least one ballot, how many board members voted for B as their first choice? (p. 34)

3. If $A = (0, -10)$ and $B = (2, 0)$, find the point(s) C on the parabola $y = x^2$ which minimizes the area of triangle ABC. (p. 35)

4. Does there exist a continuous function $y = f(x)$, defined for all real x, whose graph intersects every non-vertical line in infinitely many points? (Note that because f is a function, its graph will intersect every vertical line in exactly one point.) (p. 36)

5. A child on a pogo stick jumps 1 foot on the first jump, 2 feet on the second jump, 4 feet on the third jump, ..., 2^{n-1} feet on the nth jump. Can the child get back to the starting point by a judicious choice of directions?   (p. 37)

6. Let $S_n = \{1, n, n^2, n^3, \ldots\}$, where n is an integer greater than 1. Find the smallest number $k = k(n)$ such that there is a number which may be expressed as a sum of k (possibly repeated) elements of S_n in more than one way (rearrangements are considered the same).   (p. 37)

7. Find all integers a for which $x^3 - x + a$ has three integer roots.   (p. 38)

8. At an outdoor concert held on a huge lawn in Wohascum Municipal Park, three speakers were set up in an equilateral triangle; the idea was that the audience would be between the speakers, and anyone at the exact center of the triangle would hear each speaker at an equal "volume" (sound level). Unfortunately, an electronic malfunction caused one of the speakers to play four times as loudly as the other two. As a result, the audience tended to move away from this speaker (with some people going beyond the original triangle). This helped considerably, because the sound level from a speaker is inversely proportional to the square of the distance to that speaker (we are assuming that the sound levels depend only on the distance to the speakers). This raised a question: Where should one sit so that each speaker could be heard at the same sound level?   (p. 39)

9. Ten (not necessarily all different) integers have the property that if all but one of them are added, the possible results (depending on which one is omitted) are: 82, 83, 84, 85, 87, 89, 90, 91, 92. (This is not a misprint; there are only nine possible results.) What are the ten integers?   (p. 41)

10. Let $\mathbf{A}$ be a 4×4 matrix such that each entry of $\mathbf{A}$ is either 2 or -1. Let $d = \det(\mathbf{A})$; clearly, d is an integer. Show that d is divisible by 27.   (p. 41)

11. Consider the $n \times n$ array whose entry in the ith row, jth column is $i+j-1$. What is the smallest product of n numbers from this array, with one coming from each row and one from each column?   (p. 42)

12. Find $\int (x^6 + x^3) \sqrt[3]{x^3 + 2} \, dx.$ (p. 42)

13. Hidden among the fields of Wohascum County are some missile silos, and it recently came to light that they had attracted the attention of an invidious, but somewhat inept, foreign agent living nearby in the guise of a solid citizen. Soon the agent had two microfilms to hide, and he decided to hide them in two dark squares of a chessboard—the squares being opposite each other, and

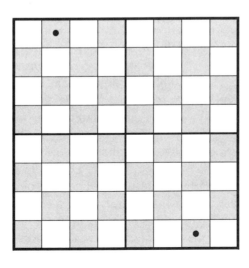

adjacent to opposite corners, as shown in the diagram. The chessboard was of a cheap collapsible variety which folded along both center lines. The agent's four-year-old daughter explored this possibility so often that the board came apart at the folds, and then she played with the four resulting pieces so they ended up in a random configuration. Her mother, unaware of the microfilms, then glued the pieces back together to form a chessboard which appeared just like the original (if somewhat the worse for wear). Considerable complications resulted when the agent could not find the films. How likely would it have been that the agent could have found them, that is, that the two squares in which they were hidden would have ended up adjacent to opposite corners?   (p. 44)

14. Let $f(x)$ be a positive, continuously differentiable function, defined for all real numbers, whose derivative is always negative. For any real number x_0, will the sequence (x_n) obtained by Newton's method $(x_{n+1} = x_n - f(x_n)/f'(x_n))$ always have limit ∞?   (p. 45)

15. Find all solutions in nonnegative integers to the system of equations

$$3x^2 - 2y^2 - 4z^2 + 54 = 0, \qquad 5x^2 - 3y^2 - 7z^2 + 74 = 0. \qquad \text{(p. 46)}$$

16. If n is a positive integer, how many real solutions are there, as a function of n, to $e^x = x^n$?   (p. 46)

17. Does there exist a positive integer whose prime factors include at most the primes 2, 3, 5, and 7 and which ends in the digits 11? If so, find the smallest such positive integer; if not, show why none exists.   (p. 47)

18. The Wohascum County Fish and Game Department issues four types of licenses, for deer, grouse, fish, and wild turkey; anyone can purchase any combination of licenses. In a recent year, (exactly) half the people who bought a grouse license also bought a turkey license. Half the people who bought a turkey license also bought a deer license. Half the people who bought a fish license also bought a grouse license, and one more than half the people who bought a fish license also bought a deer license. One third of the people who bought a deer license also bought both grouse and fish licenses. Of the people who bought deer licenses, the same number bought a grouse license as bought a fish license; a similar statement was true of buyers of turkey licenses. Anyone who bought both a grouse and a fish license also bought either a deer or a turkey license, and of these people the same number bought a deer license as bought a turkey license. Anyone who bought both a deer and a turkey li-

cense either bought both a grouse and a fish license or neither. The number of people buying a turkey license was equal to the number of people who bought some license but not a fish license. The number of people buying a grouse license was equal to the number of people buying some license but not a turkey license. The number of deer licenses sold was one more than the number of grouse licenses sold. Twelve people bought either a grouse or a deer license (or both). How many people in all bought licenses? How many licenses in all were sold?   (p. 48)

19. Given three lines in the plane which form a triangle (that is, every pair of the lines intersects, and the three intersection points are distinct), what is the set of points for which the sum of the distances to the three lines is as small as possible? (Be careful not to overlook special cases.)   (p. 50)

20. Suppose the plane $x + 2y + 3z = 0$ is a perfectly reflecting mirror. Suppose a ray of light shines down the positive x-axis and reflects off the mirror. Find the direction of the reflected ray. (Assume the law of optics which asserts that the angle of incidence equals the angle of reflection.)   (p. 51)

21.   Find the set of all solutions to

$$x^{y/z} = y^{z/x} = z^{x/y},$$

with x, y, and z positive real numbers.   (p. 52)

22.   Find all perfect squares whose base 9 representation consists only of ones.   (p. 53)

23.   The following is an excerpt from a recent article in the *Wohascum Times*. (Names have been replaced by letters.) "Because of the recent thaw, the trail for the annual Wohascum Snowmobile Race was in extremely poor condition, and it was impossible for more than two competitors to be abreast each other anywhere on the trail. Nevertheless, there was frequent passing. ... After a few miles A pulled ahead of the pack, closely followed by B and C in that order, and thereafter these three did not relinquish the top three positions in the field. However, the lead subsequently changed hands nine times among these three; meanwhile, on eight different occasions the vehicles that were running second and third at the times changed places. ... At the end of the race, C complained that B had driven recklessly just before the finish line to keep C, who was immediately behind B at the finish, from passing. ..." Can this article

be accurate? If so, can you deduce who won the race? If the article cannot be accurate, why not? (p. 53)

24. At a recent trade fair in Wohascum Center, an inventor showed a device called a "trisector," with which any straight line segment can be divided into three equal parts. The following dialogue ensued. Customer: "But I need to find the midpoint of a segment, not the points 1/3 and 2/3 of the way from one end of the segment to the other!" Inventor: "Sorry, I hadn't realized there was a market for that. I'll guess that you'll have to get some compasses and use the usual construction." Show that the inventor was wrong, that is, show how to construct the midpoint of any given segment using only a straightedge (but no compasses) and the "trisector." (p. 54)

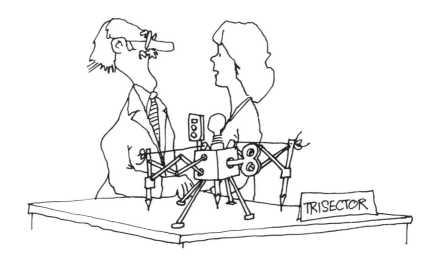

25. Consider a 12×12 chessboard (consisting of 144 1×1 squares). If one removes 3 corners, can the remainder be covered by 47 1×3 tiles? (p. 56)

26. Is there a function f, differentiable for all real x, such that

$$|f(x)| < 2 \quad \text{and} \quad f(x)f'(x) \geq \sin x?$$ (p. 57)

27. Let N be the largest possible number that can be obtained by combining the digits 1, 2, 3, and 4 using the operations addition, multiplication, and exponentiation, if the digits can be used only once. Operations can be used repeatedly, parentheses can be used, and digits can be juxtaposed (put next to each other). For instance, 12^{34}, $1 + (2 \times 3 \times 4)$, and $2^{31 \times 4}$ are all candidates, but none of these numbers is actually as large as possible. Find N. (All numbers are to be construed in base ten.)   (p. 58)

28. Babe Ruth's batting performance in the 1921 baseball season is often considered the best in the history of the game. In home games, his batting average was .404; in away games it was .354. Furthermore, his slugging percentage at home was a whopping .929, while in away games it was .772. This was based on a season total of 44 doubles, 16 triples, and 59 home runs. He had 30 more at bats in away games than in home games. What were his overall batting average and his slugging percentage for the year? (Batting average is defined to be the number of hits divided by the number of at bats. One way of defining slugging percentage is the number of hits plus the number of doubles plus twice the number of triples plus three times the number of home runs, all divided by the number of at bats. Both of these percentages are rounded to three decimal places.)   (p. 59)

29. Let C be a circle with center O, and Q a point inside C different from O. Where should a point P be located on the circumference of C to maximize $\angle OPQ$?   (p. 60)

30. Find all real solutions of the equation $\sin(\cos x) = \cos(\sin x)$.   (p. 61)

31. For a natural number $n \geq 2$, let $0 < x_1 \leq x_2 \leq \cdots \leq x_n$ be real numbers whose sum is 1. If $x_n \leq 2/3$, prove that there is some k, $1 \leq k \leq n$, for which $1/3 \leq \sum_{j=1}^{k} x_j < 2/3$.   (p. 62)

32. Is there a cubic curve $y = ax^3 + bx^2 + cx + d$, $a \neq 0$, for which the tangent lines at two distinct points coincide?   (p. 62)

33. Digital watches have become the norm even in Wohascum County. Recently three friends there were comparing their watches and found them reasonably well synchronized. In fact, all three watches were perfectly accurate (those amazing silicon chips!) in the sense that the length of a second was the same according to each watch. The time indicated shifted from one watch to the next, but in such a way that any two watches would show the same time in

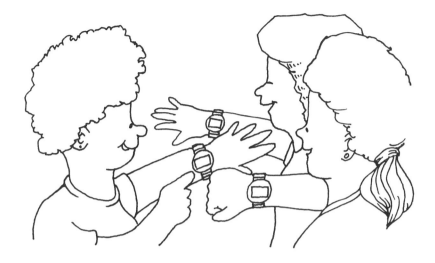

minutes for part of each minute. (A different pair of watches might show the same time in minutes for a different part of the minute.)

a. Show that there was at least one pair of watches that showed the same time in minutes for more than half of each minute.

b. Suppose there were n watches, rather than three, such that once again any two watches would show the same time in minutes for part of each minute. Find the largest number x such that at least one pair of watches necessarily showed the same time in minutes for more than the fraction x of each minute. (p. 63)

34. Let C be a circle with center O, and Q a point inside C different from O. Show that the area enclosed by the locus of the centroid of triangle OPQ as P moves about the circumference of C is independent of Q. (p. 64)

35. Describe the set of points (x, y) in the plane for which

$$\sin (x + y) = \sin x + \sin y. \qquad \text{(p. 65)}$$

36. It is shown early on in most linear algebra courses that every invertible matrix can be written as a product of elementary matrices (or, equivalently, that every invertible matrix can be reduced to the identity matrix by a finite number of row reduction steps). Show that every 2×2 matrix of determinant 1 is the product of *three* elementary matrices. (2×2 elementary matrices are matrices of types

$$\begin{pmatrix} 1 & x \\ 0 & 1 \end{pmatrix}, \quad \begin{pmatrix} 1 & 0 \\ x & 1 \end{pmatrix}, \quad \begin{pmatrix} 0 & 1 \\ 1 & 0 \end{pmatrix}, \quad \begin{pmatrix} y & 0 \\ 0 & 1 \end{pmatrix}, \quad \begin{pmatrix} 1 & 0 \\ 0 & y \end{pmatrix},$$

where $y \neq 0$ and x are arbitrary. The standard row reduction of $\begin{pmatrix} a & b \\ c & d \end{pmatrix}$ would usually use *four* row reduction steps.)   (p. 67)

37. Let $ABCD$ be a convex quadrilateral (a four-sided figure with angles less than $180°$). Find a necessary and sufficient condition for a point P to exist inside $ABCD$ such that the four triangles ABP, BCP, CDP, DAP all have the same area.   (p. 68)

38. Every week, the Wohascum Folk Dancers meet in the high school auditorium. Attendance varies, but since the dancers come in couples, there is always an even number n of dancers. In one of the dances, the dancers are in a circle; they start with the two dancers in each couple directly opposite each other. Then two dancers who are next to each other change places while all others stay in the same place; this is repeated with different pairs of adjacent dancers until, in the ending position, the two dancers in each couple are once again opposite each other, but in the opposite of the starting position (that is, every dancer is halfway around the circle from her/his original position). What is the least number of interchanges (of two adjacent dancers) necessary to do this?
                                      (p. 70)

39. Let L be a line in the plane; let A and B be points on L which are a distance 2 apart. If C is any point in the plane, there may or may not (depending on C) be a point X on the line L for which the distance from X to C is equal to the average of the distances from X to A and B. Give a precise description of the set of all points C in the plane for which there is no such point X on the line.   (p. 71)

40. Show that there exists a positive number λ such that

$$\int_0^\pi x^\lambda \sin x \, dx = 3.$$ (p. 74)

41. What is the fifth digit from the end (the ten thousands digit) of the number $5^{5^{5^{5^5}}}$?   (p. 76)

42. Describe the set of all points P in the plane such that exactly two tangent lines to the curve $y = x^3$ pass through P. (p. 77)

43. Show that if $p(x)$ is a polynomial of odd degree greater than 1, then through any point P in the plane, there will be at least one tangent line to the curve $y = p(x)$. Is this still true if $p(x)$ is of even degree? (p. 78)

44. Find a solution to the system of simultaneous equations

$$\begin{cases} x^4 - 6x^2y^2 + y^4 = 1 \\ 4x^3y - 4xy^3 = 1, \end{cases}$$

where x and y are real numbers. (p. 79)

45. Call a convex pentagon (five-sided figure with angles less than $180°$) "parallel" if each diagonal is parallel to the side with which it does not have a vertex in common. That is, $ABCDE$ is parallel if the diagonal AC is parallel to the side DE and similarly for the other four diagonals. It is easy to see that a regular pentagon is parallel, but is a parallel pentagon necessarily regular? (p. 80)

46. Find the sum of the infinite series

$$\sum_{n=1}^{\infty} \frac{1}{2n^2 - n} = 1 + \frac{1}{6} + \frac{1}{15} + \frac{1}{28} + \cdots.$$ (p. 82)

47. For any vector $\mathbf{v} = (x_1, \ldots, x_n)$ in $\mathbf{R}^n$ and any permutation σ of $1, 2, \ldots, n$, define $\sigma(\mathbf{v}) = (x_{\sigma(1)}, \ldots, x_{\sigma(n)})$. Now fix $\mathbf{v}$ and let V be the span of

$$\{\sigma(\mathbf{v}) \mid \sigma \text{ is a permutation of } 1, 2, \ldots, n\}.$$

What are the possibilities for the dimension of V? (p. 84)

48. Suppose three circles, each of radius 1, go through the same point in the plane. Let A be the set of points which lie inside at least two of the circles. What is the smallest area A can have? (p. 85)

49. How many real solutions does the equation

$$\sqrt[7]{x} - \sqrt[5]{x} = \sqrt[3]{x} - \sqrt{x}$$

have? (p. 88)

50. Let $\mathbf{A} \neq \mathbf{0}$ and $\mathbf{B}_1, \mathbf{B}_2, \mathbf{B}_3, \mathbf{B}_4$ be 2×2 matrices (with real entries) such that

$$\det(\mathbf{A} + \mathbf{B}_i) = \det \mathbf{A} + \det \mathbf{B}_i \qquad \text{for } i = 1, 2, 3, 4.$$

Show that there exist real numbers k_1, k_2, k_3, k_4, not all zero, such that

$$k_1 \mathbf{B}_1 + k_2 \mathbf{B}_2 + k_3 \mathbf{B}_3 + k_4 \mathbf{B}_4 = \mathbf{0}.$$

(**0** is the zero matrix, all of whose entries are 0.)   (p. 90)

51. As you might expect, ice fishing is a popular "outdoor" pastime during the long Wohascum County winters. Recently two ice fishermen arrived at Round Lake, which is perfectly circular, and set up their ice houses in exactly opposite directions from the center, two-thirds of the way from the center to the lakeshore. The point of this symmetrical arrangement was that any fish that could be lured would (perhaps) swim toward the closest lure, so that both fishermen would have equal expectations of their catch. Some time later, a third fisherman showed up, and since the first two adamantly refused to move their ice houses, the following problem arose. Could a third ice house be put on the lake in such a way that all three fishermen would have equal expectations

at least to the extent that the three regions, each consisting of all points on the lake for which one of the three ice houses was closest, would all have the same area? (p. 90)

52. Note that the integers $2, -3$, and 5 have the property that the difference of any two of them is an integer times the third:

$$2 - (-3) = 1 \times 5, \qquad (-3) - 5 = (-4) \times 2, \qquad 5 - 2 = (-1) \times (-3).$$

Suppose three distinct integers a, b, c have this property.

a. Show that a, b, c cannot all be positive.

b. Now suppose that a, b, c, in addition to having the above property, have no common factors (except $1, -1$). (For example, $20, -30, 50$ would not qualify, because although they have the above property, they have the common factor 10.) Is it true that one of the three integers has to be either $1, 2, -1$, or -2? (p. 92)

53. Let $\mathbf{A}$ be an $m \times n$ matrix with every entry either 0 or 1. How many such matrices $\mathbf{A}$ are there for which the number of 1's in each row and each column is even? (p. 93)

54. For three points P, Q, and R in $\mathbf{R}^3$ (or, more generally, in $\mathbf{R}^n$) we say that R is *between* P and Q if R is on the line segment connecting P and Q ($R = P$ and $R = Q$ are allowed). A subset A of $\mathbf{R}^3$ is called *convex* if for any two points P and Q in A, every point R which is between P and Q is also in A. For instance, an ellipsoid is convex, a banana is not. Now for the problem: Suppose A and B are convex subsets of $\mathbf{R}^3$. Let C be the set of all points R for which there are points P in A and Q in B such that R lies between P and Q. Does C have to be convex? (p. 95)

55. Suppose we have a configuration (set) of finitely many points in the plane which are not all on the same line. We call a point in the plane a *center* for the configuration if for every line through that point, there is an equal number of points of the configuration on either side of the line.

a. Give a necessary and sufficient condition for a configuration of four points to have a center.

b. Is it possible for a finite configuration of points (not all on the same line) to have more than one center? (p. 96)

56. Find all real solutions x of the equation

$$x^{10} - x^8 + 8x^6 - 24x^4 + 32x^2 - 48 = 0. \qquad \text{(p. 97)}$$

57. The proprietor of the Wohascum Puzzle, Game and Computer Den, a small and struggling but interesting enterprise in Wohascum Center, recently was trying to design a novel set of dice. An ordinary die, of course, is cubical, with each face showing one of the numbers 1, 2, 3, 4, 5, 6. Since each face borders on four other faces, each number is "surrounded" by four of the other numbers. The proprietor's plan was to have each die in the shape of a regular dodecahedron (with twelve pentagonal faces). Each of the numbers 1, 2, 3, 4, 5, 6 would occur on two different faces and be "surrounded" both times by all five other numbers. Is this possible? If so, in how many essentially different ways can it be done? (Two ways are considered essentially the same if one can be obtained from the other by rotating the dodecahedron.) (p. 98)

58. Let k be a positive integer. Find the largest power of 3 which divides $10^k - 1$. (p. 100)

59. Consider an arbitrary circle of radius 2 in the coordinate plane. Let n be the number of lattice points (points whose coordinates are both integers) inside, but not on, the circle.
a. What is the smallest possible value for n?
b. What is the largest possible value for n? (p. 101)

60. Let a and b be nonzero real numbers and (x_n) and (y_n) be sequences of real numbers. Given that

$$\lim_{n \to \infty} \frac{ax_n + by_n}{\sqrt{x_n^2 + y_n^2}} = 0$$

and that x_n is never 0, show that

$$\lim_{n \to \infty} \frac{y_n}{x_n}$$

exists and find its value. (p. 102)

61. The MAA Student Chapter at Wohascum College is about to organize an icosahedron-building party. Each participant will be provided twenty congruent equilateral triangles cut from old ceiling tiles. The edges of the triangles are to be beveled so they will fit together at the correct angle to form a regular icosahedron (the MAA logo). What is this angle (between adjacent faces of the icosahedron)? (p. 104)

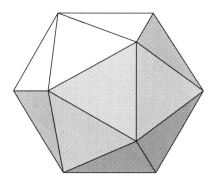

62. Given a constant C, find all functions f such that

$$f(x) + C\,f(2 - x) = (x - 1)^3 \qquad \text{for all } x. \qquad \text{(p. 105)}$$

63. The Wohascum Center branch of Wohascum National Bank recently installed a digital time/temperature display which flashes back and forth between time, temperature in degrees Fahrenheit, and temperature in degrees Centigrade (Celsius). Recently one of the local college mathematics professors became concerned when she walked by the bank and saw readings of $21°$C and $71°$F, especially since she had just taught her precocious five-year-old that same day to convert from degrees C to degrees F by multiplying by $9/5$ and adding 32 (which yields $21°$C $= 69.8°$F, which should be rounded to $70°$F). However, a bank officer explained that both readings were correct; the apparent error was due to the fact that the display device converts before rounding either Fahrenheit or Centigrade temperature to a whole number. (Thus, for example, $21.4°$C $= 70.52°$F.) Suppose that over the course of a week in spring, the temperatures measured are between $15°$C and $25°$C and that they are randomly and uniformly distributed over that interval. What is the probability that at any given time the display will appear to be in error for the reason above,

that is, that the rounded value in degrees F of the converted temperature is not the same as the value obtained by first rounding the temperature in degrees C, then converting to degrees F and rounding once more? (p. 107)

64. Sketch the set of points (x, y) in the plane which satisfy

$$(x^2 - y^2)^{2/3} + (2xy)^{2/3} = (x^2 + y^2)^{1/3}.$$ (p. 109)

65. Find all integer solutions to $x^2 + 615 = 2^n$. (p. 111)

66. Sum the infinite series

$$\sum_{n=1}^{\infty} \sin \frac{2\alpha}{3^n} \sin \frac{\alpha}{3^n}.$$ (p. 111)

67. Do there exist five rays emanating from the origin in $\mathbf{R}^3$ such that the angle between any two of these rays is obtuse (greater than a right angle)? (p. 112)

68. Find all twice continuously differentiable functions f for which there exists a constant c such that, for all real numbers a and b,

$$\left| \int_a^b f(x)\, dx - \frac{b-a}{2} \Big(f(b) + f(a) \Big) \right| \le c(b-a)^4. \qquad \text{(p. 112)}$$

69. The proprietor of the Wohascum Puzzle, Game, and Computer Den has invented a new two-person game, in which players take turns coloring edges of a cube. Three colors (red, green, and yellow) are available. The cube starts off with all edges uncolored; once an edge is colored, it cannot be colored again. Two edges with a common vertex are not allowed to have the same color. The last player to be able to color an edge wins the game.

a. Given best play on both sides, should the first or the second player win? What is the winning strategy?
b. Since there are twelve edges in all, a game can last at most twelve turns (whether or not the players use optimal strategies), and it is not hard to see that twelve turns are possible. How many twelve-turn end positions are essentially different? (Two positions are considered essentially the same if one can be obtained from the other by rotating the cube.) (p. 114)

70.   Fifty-two is the sum of two squares;
And three less is a square! So who cares?
You may think it's curious,
Perhaps it is spurious,
Are there other such numbers somewheres?

Are there other solutions in integers? If so, how many?   (p. 115)

71. Starting with a positive number $x_0 = a$, let $(x_n)_{n \geq 0}$ be the sequence of numbers such that

$$x_{n+1} = \begin{cases} x_n^2 + 1 & \text{if } n \text{ is even,} \\ \sqrt{x_n} - 1 & \text{if } n \text{ is odd.} \end{cases}$$

For what positive numbers a will there be terms of the sequence arbitrarily close to 0?   (p. 117)

72.   a. Find all positive numbers T for which

$$\int_0^T x^{-\ln x} dx = \int_T^\infty x^{-\ln x} dx.$$

b. Evaluate the above integrals for all such T, given that

$$\int_0^\infty e^{-x^2} dx = \frac{\sqrt{\pi}}{2}.$$ (p. 117)

73.   Let $f(x, y) = x^2 + y^2$ and $g(x, y) = x^2 - y^2$. Are there differentiable functions $F(z), G(z)$, and $z = h(x, y)$ such that

$$f(x, y) = F(z) \quad \text{and} \quad g(x, y) = G(z)\,?$$ (p. 118)

74.   A point $P = (a, b)$ in the plane is *rational* if both a and b are rational numbers. Find all rational points P such that the distance between P and every rational point on the line $y = 13x$ is a rational number.   (p. 119)

75.   For what positive numbers x does the series

$$\sum_{n=1}^\infty (1 - \sqrt[n]{x}) = (1 - x) + (1 - \sqrt{x}) + (1 - \sqrt[3]{x}) + \cdots$$

converge?   (p. 122)

76. Let R be a commutative ring with at least one, but only finitely many, (nonzero) zero divisors. Prove that R is finite. (p. 124)

77. Let $(a_n)_{n\geq 0}$ be a sequence of positive integers such that $a_{n+1} = 2a_n + 1$. Is there an a_0 such that the sequence consists only of prime numbers? (p. 125)

78. Suppose $c > 0$ and $0 < x_1 < x_0 < 1/c$. Suppose also that

$$x_{n+1} = cx_nx_{n-1} \qquad \text{for } n = 1, 2, \ldots.$$

a. Prove that

$$\lim_{n\to\infty} x_n = 0.$$

b. Let $\phi = (1 + \sqrt{5})/2$. Prove that

$$\lim_{n\to\infty} \frac{x_{n+1}}{x_n^\phi}$$

exists, and find it. (p. 125)

79. Given 64 points in the plane which are positioned so that 2001, but no more, distinct lines can be drawn through pairs of points, prove that at least four of the points are collinear. (p. 126)

80. Let $f_1, f_2, \ldots, f_n$ be linearly independent, differentiable functions. Prove that some $n-1$ of their derivatives $f_1', f_2', \ldots, f_n'$ are linearly independent. (p. 127)

81. Find all real numbers A and B such that

$$\left| \int_1^x \frac{1}{1+t^2} dt - A - \frac{B}{x} \right| < \frac{1}{3x^3}$$

for all $x > 1$. (p. 128)

82. The figure on the next page shows a closed knight's tour of the chessboard which is symmetric under a $180°$ rotation of the board. (A closed knight's tour is a sequence of consecutive knight moves that visits each square of the chessboard exactly once and returns to the starting point.)
a. Prove that there is no closed knight's tour of the chessboard which is symmetric under a reflection in one of the main diagonals of the board.

b. Prove that there is no closed knight's tour of the chessboard which is symmetric under a reflection in the horizontal axis through the center of the board.   (p. 129)

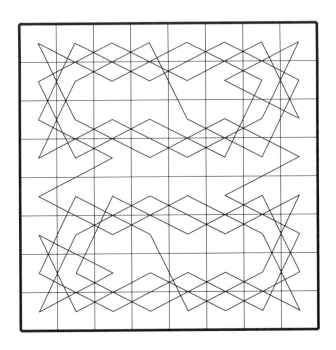

83. a. Find a sequence (a_n), $a_n > 0$, such that

$$\sum_{n=1}^{\infty} \frac{a_n}{n^3} \quad \text{and} \quad \sum_{n=1}^{\infty} \frac{1}{a_n}$$

both converge.

b. Prove that there is no sequence (a_n), $a_n > 0$, such that

$$\sum_{n=1}^{\infty} \frac{a_n}{n^2} \quad \text{and} \quad \sum_{n=1}^{\infty} \frac{1}{a_n}$$

both converge.   (p. 131)

84. It is a standard result that the limit of the indeterminate form x^x, as x approaches zero from above, is 1. What is the limit of the repeated power $x^{x^{\cdot^{\cdot^{x}}}}$ with n occurrences of x, as x approaches zero from above?   (p. 132)

85. A new subdivision is being laid out on the outskirts of Wohascum Center. There are ten north–south streets and six east–west streets, forming blocks which are exactly square. The Town Council has ordered that fire hydrants be installed at some of the intersections, in such a way that no intersection will be more than two "blocks" (really sides of blocks) away from an intersection with a hydrant. (Thus, no house will be more than $2\frac{1}{2}$ blocks from a hydrant. The blocks need not be in the same direction.) What is the smallest number of hydrants that could be used?　　　　　(p. 134)

86. Consider a triangle ABC whose angles α, β, and γ (at A, B, C respectively) satisfy $\alpha \leq \beta \leq \gamma$. Under what conditions on α, β, and γ can a beam of light placed at C be aimed at the segment AB, reflect to the segment BC, and then reflect to the vertex A? (Assume that the angle of incidence of a beam of light equals the angle of reflection.)　　　　　(p. 136)

87. Consider the following connect-the-dots game played on an $m \times n$ rectangular lattice. Two players take alternate turns, a turn consisting of drawing a line segment between two points of the lattice so that the interior of the segment intersects neither lattice points nor previously drawn segments. The last

player to be able to play wins the game. Which player has the advantage, and
what is the winning strategy?   (p. 137)

88. Find

$$\lim_{n\to\infty} \left(\sum_{k=1}^{n} \frac{1}{\binom{n}{k}}\right)^n,$$

or show that this limit does not exist.   (p. 138)

89. A person starts at the origin and makes a sequence of moves along the
real line, with the kth move being a change of $\pm k$.

a. Prove that the person can reach any integer in this way.

b. If $m(n)$ is the least number of moves required to reach a positive integer n,
prove that

$$\lim_{n\to\infty} \frac{m(n)}{\sqrt{n}}$$

exists and evaluate this limit.   (p. 139)

90. Let $N > 1$ be a positive integer and consider the functions $\varepsilon: \mathbf{Z} \to \{1, -1\}$
having period N. For what N does there exist an infinite series $\sum_{n=1}^{\infty} a_n$ with
the following properties: $\sum_{n=1}^{\infty} a_n$ diverges, whereas $\sum_{n=1}^{\infty} \varepsilon(n)a_n$ converges
for all nonconstant ε (of period N)?   (p. 141)

91. Suppose f is a continuous, increasing, bounded, real-valued function, defined on $[0, \infty)$, such that $f(0) = 0$ and $f'(0)$ exists. Show that there exists $b > 0$
for which the volume obtained by rotating the area under f from 0 to b about
the x-axis is half that of the cylinder obtained by rotating $y = f(b), 0 \le x \le b$,
about the x-axis.   (p. 141)

92. Does the Maclaurin series (Taylor series at 0) for e^{x-x^3} have any zero
coefficients?   (p. 142)

93. Let a and d be relatively prime positive integers, and consider the sequence $a, a+d, a+4d, a+9d, \ldots, a+n^2d, \ldots$. Given a positive integer b, can
one always find an integer in the sequence which is relatively prime to b?
                                                                         (p. 144)

94. Find the smallest possible n for which there exist integers $x_1, x_2, \ldots, x_n$
such that each integer between 1000 and 2000 (inclusive) can be written as the

sum, without repetition, of one or more of the integers $x_1, x_2, \ldots, x_n$. (It is not required that all such sums lie between 1000 and 2000, just that any integer between 1000 and 2000 be such a sum.) (p. 144)

95. Define $(x_n)_{n \geq 1}$ by

$$x_1 = 1, \qquad x_{n+1} = \frac{1}{\sqrt{2}}\sqrt{1 - \sqrt{1 - x_n^2}}.$$

a. Show that

$$\lim_{n \to \infty} x_n$$

exists and find this limit.

b. Show that there is a unique number A for which

$$L = \lim_{n \to \infty} \frac{x_n}{A^n}$$

exists as a finite nonzero number. Evaluate L for this value of A. (p. 145)

96. Consider the line segments in the xy-plane formed by connecting points on the positive x-axis with x an integer to points on the positive y-axis with y an integer. We call a point in the first quadrant an *I-point* if it is the intersection of two such line segments. We call a point an *L-point* if there is a sequence of distinct I-points whose limit is the given point. Prove or disprove: If (x, y) is an L-point, then either x or y (or both) is an integer. (p. 146)

97. a. Find all lines which are tangent to both of the parabolas

$$y = x^2 \qquad \text{and} \qquad y = -x^2 + 4x - 4.$$

b. Now suppose $f(x)$ and $g(x)$ are any two quadratic polynomials. Find geometric criteria that determine the number of lines tangent to both of the parabolas $y = f(x)$ and $y = g(x)$. (p. 148)

98. Suppose we are given an m-gon (polygon with m sides, and including the interior for our purposes) and an n-gon in the plane. Consider their intersection; assume this intersection is itself a polygon (other possibilities would include the intersection being empty or consisting of a line segment).

a. If the m-gon and the n-gon are convex, what is the maximal number of sides their intersection can have?

b. Is the result from (a) still correct if only one of the polygons is assumed to be convex?

(Note: A subset of the plane is *convex* if for every two points of the subset, every point of the line segment between them is also in the subset. In particular, a polygon is convex if each of its interior angles is less than 180°.)   (p. 151)

99.   Every year, the first warm days of summer tempt Lake Wohascum's citizens to venture out into the local parks; in fact, one day last May, the MAA Student Chapter held an impromptu picnic. A few insects were out as well, and at one point an insect dropped from a tree onto a paper plate (fortunately an empty one) and crawled off. Although this did not rank with Newton's apple as a source of inspiration, it did lead the club to wonder: If an insect starts at a random point inside a circle of radius R and crawls in a straight line in a random direction until it reaches the edge of the circle, what will be the average distance it travels to the perimeter of the circle? ("Random point" means that given two equal areas within the circle, the insect is equally likely to start in one as in the other; "random direction" means that given two equal angles with vertex at the point, the insect is equally likely to crawl off inside one as the other.)   (p. 154)

100.   Let $ABCD$ be a parallelogram in the plane. Describe and sketch the set of all points P in the plane for which there is an ellipse with the property that the points A, B, C, D, and P all lie on the ellipse.   (p. 156)

101. Let x_0 be a rational number, and let $(x_n)_{n\geq 0}$ be the sequence defined recursively by

$$x_{n+1} = \left| \frac{2x_n^3}{3x_n^2 - 4} \right|.$$

Prove that this sequence converges, and find its limit as a function of x_0.

(p. 159)

102. Let f be a continuous function on $[0, 1]$, which is bounded below by 1, but is not identically 1. Let R be the region in the plane given by $0 \leq x \leq 1$, $1 \leq y \leq f(x)$. Let

$$R_1 = \{(x, y) \in R \mid y \leq \bar{y}\} \quad \text{and} \quad R_2 = \{(x, y) \in R \mid y \geq \bar{y}\},$$

where $\bar{y}$ is the y-coordinate of the centroid of R. Can the volume obtained by rotating R_1 about the x-axis equal that obtained by rotating R_2 about the x-axis?

(p. 161)

103. Let $n \geq 3$ be a positive integer. Begin with a circle with n marks about it. Starting at a given point on the circle, move clockwise, skipping over the next two marks and placing a new mark; the circle now has $n+1$ marks. Repeat the procedure beginning at the new mark. Must a mark eventually appear between each pair of the original marks?

(p. 163)

104. Let

$$c = \sum_{n=1}^{\infty} \frac{1}{n(2^n - 1)} = 1 + \frac{1}{6} + \frac{1}{21} + \frac{1}{60} + \cdots.$$

Show that

$$e^c = \frac{2}{1} \cdot \frac{4}{3} \cdot \frac{8}{7} \cdot \frac{16}{15} \cdot \cdots.$$

(p. 164)

105. Let $q(x) = x^2 + ax + b$ be a quadratic polynomial with real roots. Must all roots of $p(x) = x^3 + ax^2 + (b - 3)x - a$ be real?

(p. 166)

106. Let $p(x) = x^3 + a_1x^2 + a_2x + a_3$ have rational coefficients and have roots r_1, r_2, r_3. If $r_1 - r_2$ is rational, must $r_1, r_2,$ and r_3 be rational? (p. 167)

107. Let $f(x) = x^3 - 3x + 3$. Prove that for any positive integer P, there is a "seed" value x_0 such that the sequence $x_0, x_1, x_2, \ldots$ obtained from Newton's

method, given by

$$x_{n+1} = x_n - \frac{f(x_n)}{f'(x_n)},$$

has period P.   (p. 169)

108.   Show that

$$\sum_{k=0}^{n} \frac{(-1)^k}{2n + 2k + 1} \binom{n}{k} = \frac{\left(2^n (2n)!\right)^2}{(4n + 1)!}.$$ (p. 170)

109.   Suppose a and b are distinct real numbers such that

$$a - b, \ a^2 - b^2, \ldots, a^k - b^k, \ldots$$

are all integers.

a. Must a and b be rational?

b. Must a and b be integers?   (p. 173)

110.   The mayor of Wohascum Center has ten pairs of dress socks, ranging through ten shades of color from medium gray (1) to black (10). When he has worn all ten pairs, the socks are washed and dried together. Unfortunately, the light in the laundry room is very poor and all the socks look black there; thus, the socks get paired at random after they are removed from the drier. A pair of socks is unacceptable for wearing if the colors of the two socks differ by more than one shade.

What is the probability that the socks will be paired in such a way that all ten pairs are acceptable? (p. 174)

111. Let $p(x, y)$ be a real polynomial.

a. If $p(x, y) = 0$ for infinitely many (x, y) on the unit circle $x^2 + y^2 = 1$, must $p(x, y) = 0$ on the unit circle?

b. If $p(x, y) = 0$ on the unit circle, is $p(x, y)$ necessarily divisible by $x^2 + y^2 - 1$? (p. 176)

112. Find all real polynomials $p(x)$, whose roots are real, for which

$$p(x^2 - 1) = p(x)p(-x).$$ (p. 178)

113. Consider sequences of points in the plane that are obtained as follows: The first point of each sequence is the origin. The second point is reached from the first by moving one unit in any of the four "axis" directions (east, north, west, south). The third point is reached from the second by moving $1/2$ unit in any of the four axis directions (but not necessarily in the same direction), and so on. Thus, each point is reached from the previous point by moving in any of the four axis directions, and each move is half the size of the previous move. We call a point *approachable* if it is the limit of some sequence of the above type.

Describe the set of all approachable points in the plane. That is, find a necessary and sufficient condition for (x, y) to be approachable. (p. 179)

114. A gambling game is played as follows: D dollar bills are distributed in some manner among N indistinguishable envelopes, which are then mixed up in a large bag. The player buys random envelopes, one at a time, for one dollar and examines their contents as they are purchased. If the player can buy as many or as few envelopes as desired, and, furthermore, knows the initial distribution of the money, then for what distribution(s) will the player's expected net return be maximized? (p. 181)

115. Let $\alpha = .d_1 d_2 d_3 \ldots$ be a decimal representation of a real number between 0 and 1. Let r be a real number with $|r| < 1$.

a. If α and r are rational, must $\sum_{i=1}^{\infty} d_i r^i$ be rational?

b. If α and r are rational, must $\sum_{i=1}^{\infty} id_i r^i$ be rational?

c. If r and $\sum_{i=1}^{\infty} d_i r^i$ are rational, must α be rational? (p. 183)

116. Let $\mathcal{L}_1$ and $\mathcal{L}_2$ be skew lines in space (that is, straight lines which do not lie in the same plane). How many straight lines $\mathcal{L}$ have the property that every point on $\mathcal{L}$ has the same distance to $\mathcal{L}_1$ as to $\mathcal{L}_2$? (p. 186)

117. We call a sequence $(x_n)_{n\geq 1}$ a *superinteger* if (i) each x_n is a nonnegative integer less than 10^n and (ii) the last n digits of x_{n+1} form x_n. One example of such a sequence is $1, 21, 021, 1021, 21021, 021021, \ldots$, which we abbreviate by $\ldots 21021$. Note that the digit 0 is allowed (as in the example) and that (unlike the example) there may not be a pattern to the digits. The ordinary positive integers are just those superintegers with only finitely many nonzero digits. We can do arithmetic with superintegers; for instance, if x is the superinteger above, then the product xy of x with the superinteger $y = \ldots 66666$ is found as follows:

$1 \times 6 = 6$: the last digit of xy is 6.
$21 \times 66 = 1386$: the last two digits of xy are 86.
$021 \times 666 = 13986$: the last three digits of xy are 986.
$1021 \times 6666 = 6805986$: the last four digits of xy are 5986, etc.

Is it possible for two nonzero superintegers to have product $0 = \ldots 00000$? (p. 187)

118. If $\sum a_n$ converges, must there exist a periodic function $\varepsilon : \mathbf{Z} \to \{1, -1\}$ such that $\sum \varepsilon(n)|a_n|$ converges? (p. 188)

119. Let $f(x) = x - 1/x$. For any real number x_0, consider the sequence defined by $x_0, x_1 = f(x_0), \ldots, x_{n+1} = f(x_n), \ldots$, provided $x_n \neq 0$. Define x_0 to be a *T-number* if the sequence terminates, that is, if $x_n = 0$ for some n. (For example, -1 is a T-number since $f(-1) = 0$, but $\sqrt{2}$ is not, since the sequence

$$\sqrt{2}, \quad 1/\sqrt{2} = f(\sqrt{2}), \quad -1/\sqrt{2} = f(1/\sqrt{2}), \quad 1/\sqrt{2} = f(-1/\sqrt{2}), \quad \ldots$$

does not terminate.)

a. Show that the set of all T-numbers is countably infinite (denumerable).

b. Does every open interval contain a T-number? (p. 190)

120. For n a positive integer, show that the number of integral solutions (x, y) of $x^2 + xy + y^2 = n$ is finite and a multiple of 6.   (p. 192)

121. For what real numbers x can one say the following?

a. For each positive integer n, there exists an integer m such that

$$\left| x - \frac{m}{n} \right| < \frac{1}{3n}.$$

b. For each positive integer n, there exists an integer m such that

$$\left| x - \frac{m}{n} \right| \leq \frac{1}{3n}.$$ (p. 194)

122. Let $\mathbf{Z}_n$ be the set $\{0, 1, \ldots, n-1\}$ with addition modulo n. Consider subsets S_n of $\mathbf{Z}_n$ such that $(S_n + k) \cap S_n$ is nonempty for every k in $\mathbf{Z}_n$. Let $f(n)$ denote the minimal number of elements in such a subset. Find

$$\lim_{n \to \infty} \frac{\ln f(n)}{\ln n},$$

or show that this limit does not exist.   (p. 195)

123. a. If a rational function (a quotient of two real polynomials) takes on rational values for infinitely many rational numbers, prove that it may be expressed as the quotient of two polynomials with rational coefficients.

b. If a rational function takes on integral values for infinitely many integers, prove that it must be a polynomial with rational coefficients.   (p. 196)

124. Can there be a multiplicative $n \times n$ magic square $(n > 1)$ with entries $1, 2, \ldots, n^2$? That is, does there exist an integer $n > 1$ for which the numbers $1, 2, \ldots, n^2$ can be placed in a square so that the product of all the numbers in any row or column is always the same?   (p. 199)

125. Note that if the edges of a regular octahedron have length 1, then the distance between any two of its vertices is either 1 or $\sqrt{2}$. Are there other configurations of six points in $\mathbf{R}^3$ for which the distance between any two of the points is either 1 or $\sqrt{2}$? If so, find them.   (p. 200)

126. Let a and b be positive real numbers, and define a sequence (x_n) by

$$x_0 = a, \ x_1 = b, \ x_{n+1} = \frac{1}{2}\left(\frac{1}{x_n} + x_{n-1}\right).$$

a. For what values of a and b will this sequence be periodic?

b. Show that given a, there exists a unique b for which the sequence converges.

(p. 205)

127. Consider the equation $x^2 + \cos^2 x = \alpha \cos x$, where α is some positive real number.

a. For what value or values of α does the equation have a unique solution?

b. For how many values of α does the equation have precisely four solutions?

(p. 209)

128. Fast Eddie needs to double his money; he can only do so by playing a certain win-lose game, in which the probability of winning is p. However, he can play this game as many or as few times as he wishes, and in a particular game he can bet any desired fraction of his bankroll. The game pays even money (the odds are one-to-one). Assuming he follows an optimal strategy if one is available, what is the probability, as a function of p, that Fast Eddie will succeed in doubling his money?

(p. 212)

129. Define a *die* to be a convex polyhedron. For what n is there a fair die with n faces? By fair, we mean that, given any two faces, there exists a symmetry of the polyhedron which takes the first face to the second.

(p. 214)

130. Prove that

$$\det \begin{pmatrix} 1 & 4 & 9 & \cdots & n^2 \\ n^2 & 1 & 4 & \cdots & (n-1)^2 \\ \vdots & \vdots & \vdots & \vdots & \vdots \\ 4 & 9 & 16 & \cdots & 1 \end{pmatrix}$$

$$= (-1)^{n-1} \frac{n^{n-2}(n+1)(2n+1)\big((n+2)^n - n^n\big)}{12} .$$

(p. 216)

THE SOLUTIONS

Problem 1

Find all solutions in integers of $x^3 + 2y^3 = 4z^3$.

Answer. The only solution is $(x, y, z) = (0, 0, 0)$.

Solution 1. First note that if (x, y, z) is a solution in integers, then x must be even, say $x = 2w$. Substituting this, dividing by 2, and subtracting y^3, we see that $(-y)^3 + 2z^3 = 4w^3$, so $(-y, z, w) = (-y, z, x/2)$ is another solution. Now repeat this process to get (x, y, z), $(-y, z, x/2)$, $(-z, x/2, -y/2)$, $(-x/2, -y/2, -z/2)$ as successive integer solutions. Conclusion: If (x, y, z) is a solution, then so is $(-x/2, -y/2, -z/2)$, and in particular, x, y, and z are all even. But if x, y, and z were not all zero, we could keep replacing (x, y, z) by $(-x/2, -y/2, -z/2)$ and eventually arrive at a solution containing an odd integer, a contradiction.

Solution 2. Suppose (x, y, z) is a nonzero solution for which $|x|^3 + 2|y|^3 + 4|z|^3$ is minimized. Clearly x is even, say $x = 2w$. We have $8w^3 + 2y^3 = 4z^3$ or $(-y)^3 + 2z^3 = 4w^3$. Then $(-y, z, w)$ is a nonzero solution and

$$| -y|^3 + 2|z|^3 + 4|w|^3 = \frac{|x|^3 + 2|y|^3 + 4|z|^3}{2},$$

a contradiction. Therefore $(0, 0, 0)$ is the only solution.

Comment. This solution method, in which it is shown that any solution gives rise to a "smaller" solution, is known as the *method of infinite descent*. It was introduced by Pierre de Fermat (1601–1665), who used intricate cases of the method to solve diophantine equations that were beyond the reach of any of his contemporaries.

Problem 2

Each of twenty commissioners ranked three candidates (A, B, and C) in order of preference, with no abstentions. It was found that 11 commissioners preferred A to B, 12 preferred C to A, but 14 preferred B to C. Given that every possible order of A, B, and C appeared on at least one ballot, how many commissioners voted for B as their first choice?

Solution. Eight commissioners voted for B. To see this, we will use the given information to study how many voters chose each order of A, B, C.

The six orders of preference are ABC, ACB, BAC, BCA, CAB, CBA; assume they receive a, b, c, d, e, f votes respectively. We know that

$$a + b + e = 11 \qquad (A \text{ over } B) \qquad (1)$$

$$d + e + f = 12 \qquad (C \text{ over } A) \qquad (2)$$

$$a + c + d = 14 \qquad (B \text{ over } C). \qquad (3)$$

Because 20 votes were cast, we also know that

$$c + d + f = 9 \qquad (B \text{ over } A) \qquad (4)$$

$$a + b + c = 8 \qquad (A \text{ over } C) \qquad (5)$$

$$b + e + f = 6 \qquad (C \text{ over } B). \qquad (6)$$

There are many ways one might proceed. For example, equations (3) and (4) imply that $a = 5+f$, and equations (1) and (5) imply that $e = c+3$. Substituting these into (1) yields $f + b + c = 3$, and therefore $b = c = f = 1$. It follows that $a = 6$, $e = 4$, and $d = 7$.

The number of commissioners voting for B as their first choice is therefore $c + d = 1 + 7 = 8$ (A has 7 first place votes, and C has 5).

Comments. The answer to this question would have been the same had we known only that *at least* 14 commissioners preferred B over C.

The seemingly paradoxical nature of the commissioners' preferences (A preferred to B, B preferred to C, C preferred to A), an example of "nontransitive dominance," is not uncommon when individual choices are pooled.

Problem 3

If $A = (0, -10)$ and $B = (2, 0)$, find the point(s) C on the parabola $y = x^2$ which minimizes the area of triangle ABC.

Answer. The area of triangle ABC is minimized when $C = (5/2, 25/4)$.

Solution 1. The area of triangle ABC is half the product of the length of AB, which is fixed, and the length of the altitude from C to AB, which varies with C. The length of this altitude is the distance between C and the line AB. Note that the line AB, given by $y = 5x - 10$, does not intersect the parabola $y = x^2$. Thus, for $C = (x, x^2)$ on the parabola to minimize the area, the tangent line at C must be parallel to AB. This occurs when $2x = 5$, or $C = (5/2, 25/4)$.

Solution 2. (Gerald A. Heuer, Concordia College)   For a straightforward solution based on the properties of the cross product, recall that

$$\text{Area } ABC = \frac{1}{2}\left|\overrightarrow{AB} \times \overrightarrow{AC}\right|.$$

Now

$$\overrightarrow{AB} \times \overrightarrow{AC} = \det \begin{pmatrix} \mathbf{i} & \mathbf{j} & \mathbf{k} \\ 2 & 10 & 0 \\ x & x^2 + 10 & 0 \end{pmatrix} = 2(x^2 - 5x + 10)\mathbf{k}.$$

Thus, Area $ABC = x^2 - 5x + 10 = (x - 5/2)^2 + 15/4$, and this is minimized when $x = 5/2$.

Solution 3. (Les Nelson, St. Olaf College)   The transformation (a shear followed by a vertical translation)

$$\begin{pmatrix} X \\ Y \end{pmatrix} = \begin{pmatrix} 1 & 0 \\ -5 & 1 \end{pmatrix} \begin{pmatrix} x \\ y \end{pmatrix} + \begin{pmatrix} 0 \\ 10 \end{pmatrix}$$

is area-preserving because

$$\det \begin{pmatrix} 1 & 0 \\ -5 & 1 \end{pmatrix} = 1,$$

and it takes $A = (0, -10)$ to $A' = (0,0)$, $B = (2,0)$ to $B' = (2,0)$, and $C = (x, x^2)$ to $C' = (x, x^2 - 5x + 10)$. Therefore,

$$\text{Area } ABC = \text{Area } A'B'C' = x^2 - 5x + 10 = (x - 5/2)^2 + 15/4,$$

which is minimized when $x = 5/2$.

Problem 4

Does there exist a continuous function $y = f(x)$, defined for all real x, whose graph intersects every non-vertical line in infinitely many points?

Solution. Yes, there is such a function; an example is $f(x) = x^2 \sin x$. The graph of this function oscillates between the graph of $y = x^2$ (which it intersects when $\sin x = 1$, that is, when $x = \pi/2 + 2k\pi$) and the graph of $y = -x^2$ (which it intersects at $x = 3\pi/2 + 2k\pi$).

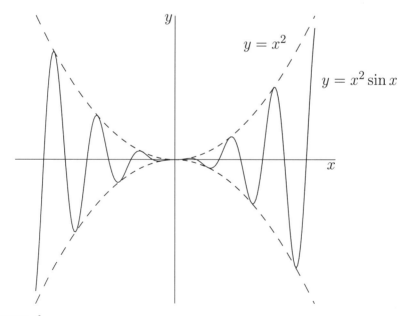

FIGURE 1

Suppose $y = mx + b$ is the equation of a non-vertical line. Since

$$\lim_{x \to \infty} \frac{mx + b}{x^2} = 0,$$

we know that for x large enough

$$\left| \frac{mx + b}{x^2} \right| < 1,$$

or equivalently $-x^2 < mx + b < x^2$. Therefore, the line $y = mx + b$ will intersect the graph $y = x^2 \sin x$ in each interval $(\pi/2 + 2k\pi, \ 3\pi/2 + 2k\pi)$ for sufficiently large integers k.

Problem 5

A child on a pogo stick jumps 1 foot on the first jump, 2 feet on the second jump, 4 feet on the third jump, ..., 2^{n-1} feet on the nth jump. Can the child get back to the starting point by a judicious choice of directions?

Solution. No. The child will always "overshoot" the starting point, because after n jumps the distance from the starting point to the child's location is at most $1 + 2 + 4 + \ldots + 2^{n-1} = 2^n - 1$ feet. Since the $(n+1)$st jump is 2^n feet, it can never return the child exactly to the starting point.

Problem 6

Let $S_n = \{1, n, n^2, n^3, \ldots\}$, where n is an integer greater than 1. Find the smallest number $k = k(n)$ such that there is a number which may be expressed as a sum of k (possibly repeated) elements of S_n in more than one way (rearrangements are considered the same).

Answer. We will show that $k(n) = n + 1$.

Solution 1. Suppose we have $a_1, a_2, \ldots, a_k$ and $b_1, b_2, \ldots, b_k$ in S_n such that

$$a_1 + a_2 + \cdots + a_k = b_1 + b_2 + \cdots + b_k,$$

where $a_i \leq a_{i+1}$, $b_i \leq b_{i+1}$, and, for some i, $a_i \neq b_i$. If k is the smallest integer for which such a sum exists, then clearly $a_i \neq b_j$ for all i, j, since otherwise we could just drop the equal terms from the sums. By symmetry, we may now assume $a_1 < b_1$. Upon dividing both sides of the equation by a_1, we may assume $a_1 = 1$. Since, at this point, every b_j is divisible by n, there must be at least n 1's on the left side of the equation. If we had $k = n$, then the right side would be strictly greater than the left. Therefore, $k \geq n + 1$. The expression

$$\underbrace{1 + 1 + \cdots + 1}_{n \text{ times}} + n^2 = \underbrace{n + n + \cdots + n}_{n+1 \text{ times}}$$

shows that $k = n + 1$.

Solution 2. First, $k(n) \leq n + 1$ because

$$\underbrace{1 + 1 + \cdots + 1}_{n \text{ times}} + n^2 = \underbrace{n + n + \cdots + n}_{n+1 \text{ times}}.$$

Next, $k(n) \geq n$. To see this, recall that every positive integer N has a *unique* base n representation; that is, there are *unique* "digits" $d_0, d_1, \ldots, d_s$, $0 \leq d_i < n$, such that

$$N = d_0 + d_1 n + d_2 n^2 + \cdots + d_s n^s$$

$$= \underbrace{1 + \cdots + 1}_{d_0} + \underbrace{n + \cdots + n}_{d_1} + \underbrace{n^2 + \cdots + n^2}_{d_2} + \cdots + \underbrace{n^s + \cdots + n^s}_{d_s}$$

This means that if N has a second representation as a sum of elements from S_n, at least one of the elements of S_n must occur at least n times. Thus $k(n) \geq n$.

Finally, $k(n) \neq n$. For suppose $k(n) = n$ and that M is an integer with two different representations,

$$M = s_1 + \cdots + s_n = t_1 + \cdots + t_n, \qquad s_i, t_i \in S_n.$$

By the uniqueness of base n representation, at least one of these representations, say the left side, is *not* the base n representation. As argued in the preceding paragraph, all of the s_i must be equal, say to n^s. Thus, $M = n^{s+1}$. But this means that the right side is not the base n representation of M either, so again, all of the t_i are equal, say to n^t. It follows that $M = n^{s+1} = n^{t+1}$ and therefore $s = t$, $s_1 = \cdots = s_n = t_1 = \cdots = t_n$. This contradicts our assumption that the representations are different.

Thus, $n < k(n) \leq n + 1$, so $k(n) = n + 1$.

Problem 7

Find all integers a for which $x^3 - x + a$ has three integer roots.

Answer. The only integer a for which $x^3 - x + a$ has three integer roots is $a = 0$. The roots are then $-1, 0, 1$.

Solution 1. The factorization $x^3 - x = (x-1)x(x+1)$ clearly implies $x^3 - x$ is strictly increasing for $x \geq 1$ and is strictly increasing for $x \leq -1$. (One could prove this using calculus, as well.) Thus, the polynomial $x^3 - x + a$ can have at most one positive integral root and one negative integral root. A third integral root must then be 0, hence $a = 0$.

Solution 2. (Suggested by students at the 1990 United States Mathematical Olympiad training program)   Let r_1, r_2, r_3 be the integral roots of $x^3 - x + a$. Writing the coefficients of the polynomial as symmetric functions of its roots, we have

$$r_1 + r_2 + r_3 = 0 \quad \text{and} \quad r_1 r_2 + r_1 r_3 + r_2 r_3 = -1.$$

Combining these yields

$$r_1^2 + r_2^2 + r_3^2 = (r_1 + r_2 + r_3)^2 - 2(r_1 r_2 + r_1 r_3 + r_2 r_3) = 2.$$

We conclude that one of the roots is 0, forcing $a = 0$, and the other two roots are -1 and 1.

Solution 3. Let r_1, r_2, r_3 be the roots of $x^3 - x + a$. The discriminant of this cubic is

$$\left((r_1 - r_2)(r_1 - r_3)(r_2 - r_3) \right)^2 = -4(-1)^3 - 27a^2 = 4 - 27a^2.$$

If all three roots are integral, then the above discriminant is a perfect square, say s^2. Then $4 = 27a^2 + s^2$, which implies $a = 0$.

Problem 8

Three loudspeakers are placed so as to form an equilateral triangle. One of the speakers plays four times as loudly as the other two. Assuming that the sound level from a speaker is inversely proportional to the square of the distance to that speaker (and that the sound levels depend only on the distance to the speakers), where should one sit so that each speaker will be heard at the same sound level?

Solution 1. The listener should sit at the reflection of the center of the triangle in the side opposite the louder speaker.

Because the sound level for a speaker is inversely proportional to the square of the distance from that speaker, we seek a point twice as far from the louder speaker as from the other two speakers. Recall that the intersection of the medians of a triangle occurs at the point two-thirds of the way from a vertex of the triangle to the midpoint of the opposite side. Furthermore, in an equilateral triangle, each median is perpendicular to its corresponding side. Thus, the reflection of the center of the triangle in the side opposite the louder speaker is twice as far from that speaker as is the center of the triangle. On the other hand, noting that the reflection of a given point in a line has the same distance to any point on the line as the given point, we see that the reflected point and the center of the triangle are equidistant from the two other speakers. We have found a point where the sound levels from the three speakers are equal.

Solution 2. Let the louder speaker be at point A, and the other two speakers be at B and C. Then the desired point is the intersection of the perpendicular to AB at B and the perpendicular bisector of BC.

We again begin with the observation that we want a point P for which $PB = PC = \frac{1}{2}PA$. The set of points equidistant from B and C is the perpendicular bisector of the line segment BC, which passes through A since $AB = AC$. If P is the intersection of the perpendicular to AB at B and the perpendicular bisector of BC, clearly $PB = PC$. In addition, the triangle APB is a $30°$–$60°$–$90°$ triangle, hence $PB = \frac{1}{2}PA$, as required.

It is now easy to go on to show that P is the only such point. Suppose Q is on the line PA. If A is between P and Q, then $QA < QB = QC$. If Q is on the same side of A as P, then $QB \geq QA \sin 30° = \frac{1}{2}QA$, with equality if and only if the lines QB and AB are perpendicular.

Solution 3. If the identical speakers are at the points $(-1, 0)$ and $(1, 0)$ in the plane, and the louder speaker is at $(0, \sqrt{3})$, then the unique point we seek is at $(0, -1/\sqrt{3})$.

Since the points $(-1, 0)$, $(1, 0)$, and $(0, \sqrt{3})$ form the vertices of an equilateral triangle, there is no loss of generality in choosing coordinates so that the speakers are at these points. By looking at the squares of the distances from a point (x, y) to these points, we arrive at the condition

$$(x + 1)^2 + y^2 = (x - 1)^2 + y^2 = \tfrac{1}{4}\left(x^2 + (y - \sqrt{3})^2\right).$$

The first equality implies $x = 0$, which we substitute into the second equality, obtaining

$$1 + y^2 = \tfrac{1}{4} \left(y^2 - 2\sqrt{3}y + 3 \right).$$

After combining terms, we find that $y = -1/\sqrt{3}$ is the only solution to this last equation. Thus, the only point where the sound levels from the three speakers are equal is $(0, -1/\sqrt{3})$.

Problem 9

Ten (not necessarily all different) integers have the property that if all but one of them are added, the possible results (depending on which one is omitted) are: $82, 83, 84, 85, 87, 89, 90, 91, 92$. What are the ten integers?

Solution. The integers are $5, 6, 7, 7, 8, 10, 12, 13, 14, 15$.

Let $n_1 \leq n_2 \leq \cdots \leq n_{10}$ be the ten integers, and let $S = n_1 + n_2 + \cdots + n_{10}$ be their sum. If all but n_1 are added, the result is $S - n_1$; similarly, if all but n_2 are added, the result is $S - n_2$; and so forth. If the ten results are added, we get $(S - n_1) + (S - n_2) + \cdots + (S - n_{10}) = 10S - S = 9S$. Now we are given that there are only nine possible different results: $82, 83, 84, 85, 87, 89,$ $90, 91, 92$. Let x denote the sum that occurs twice. Then the sum of the ten results is $82 + 83 + 84 + 85 + 87 + 89 + 90 + 91 + 92 + x$; by the above, this sum is also $9S$. Therefore, $783 + x = 9S$; equivalently, $x = 9(S - 87)$. This shows that x must be divisible by 9. However, of the given results $82, 83, \ldots, 92$, only 90 is divisible by 9, so $x = 90$, and $S = 97$. Subtracting $82, 83, \ldots, 92$, with 90 occurring twice, from 97, we find the ten integers listed above.

Problem 10

Let $\mathbf{A}$ be a 4×4 matrix such that each entry of $\mathbf{A}$ is either 2 or -1. Let $d = \det(\mathbf{A})$; clearly, d is an integer. Show that d is divisible by 27.

Solution. Let $\mathbf{B}$ be the matrix obtained from $\mathbf{A}$ by subtracting row one of $\mathbf{A}$ from each of the other three rows. Then $\det \mathbf{A} = \det \mathbf{B}$. Each entry in the last three rows of $\mathbf{B}$ is $-3, 0,$ or 3, and therefore is divisible by 3. Now let $\mathbf{C}$ be the matrix obtained from $\mathbf{B}$ by dividing all these entries (in the last three rows) by 3. Then all entries of $\mathbf{C}$ are integers, so $\det \mathbf{C}$ is an integer; moreover, $\det \mathbf{A} = \det \mathbf{B} = 3^3 \det \mathbf{C}$, so $\det \mathbf{A}$ is divisible by 27.

Problem 11

Consider the $n \times n$ array whose entry in the ith row, jth column is $i + j - 1$. What is the smallest product of n numbers from this array, with one coming from each row and one from each column?

Solution. The smallest product is $1 \cdot 3 \cdot 5 \cdots (2n-1)$. This value of the product occurs when the n numbers are on the main diagonal of the array (positions $(1, 1), (2, 2), \ldots, (n, n)$).

Because there are only finitely many possible products, there is a smallest product. We show this could not occur for any other choice of numbers.

Assume $n \geq 2$, and suppose we choose the n numbers from row i, column i', $i = 1, 2, \ldots, n$. Unless $i' = i$ for all i, there will be some i and j, $i < j$, for which $i' > j'$. If we choose entries (i, j') and (j, i') instead of (i, j) and (i', j'), keeping the other $n - 2$ entries the same, the product of the two entries will be $(i + j' - 1)(j + i' - 1)$ instead of $(i + i' - 1)(j + j' - 1)$. Since

$$(i + j' - 1)(j + i' - 1) - (i + i' - 1)(j + j' - 1) = (j - i)(j' - i') < 0,$$

the new product will be smaller, so we are done.

Problem 12

Find $\displaystyle\int (x^6 + x^3) \sqrt[3]{x^3 + 2} \, dx$.

Answer.

$$\int (x^6 + x^3) \sqrt[3]{x^3 + 2} \, dx = \frac{1}{8}(x^6 + 2x^3)^{4/3} + C = \frac{1}{8}(x^7 + 2x^4) \sqrt[3]{x^3 + 2} + C.$$

Solution 1. We start by bringing a factor x inside the cube root:

$$\int (x^6 + x^3) \sqrt[3]{x^3 + 2} \, dx = \int (x^5 + x^2) \sqrt[3]{x^6 + 2x^3} \, dx.$$

The substitution $u = x^6 + 2x^3$ now yields the answer.

Solution 2. Setting $u = x^3 + 2$, we find

$$du = 3x^2 \, dx, \qquad dx = \frac{du}{3(u-2)^{2/3}}.$$

Thus,

$$\int (x^6 + x^3)\sqrt[3]{x^3 + 2}\, dx = \int \frac{(u-2)(u-1)u^{1/3}}{3(u-2)^{2/3}} \, du$$

$$= \frac{1}{3} \int (u^2 - 2u)^{1/3}(u-1) \, du$$

$$= \frac{1}{6} \left(\frac{(u^2 - 2u)^{4/3}}{4/3} \right) + C$$

$$= \frac{1}{8} (x^6 + 2x^3)^{4/3} + C.$$

Solution 3. (Eugene Luks, University of Oregon)   Starting with the term of highest degree and integrating by parts, we get

$$\int x^6 \sqrt[3]{x^3 + 2}\, dx = \int x^4 \left(x^2 \sqrt[3]{x^3 + 2} \right) dx$$

$$= \frac{1}{4} x^4 \, (x^3 + 2)^{4/3} - \int x^3 (x^3 + 2)^{4/3} \, dx$$

$$= \frac{1}{4} x^4 (x^3 + 2)^{4/3} - \int (x^6 + 2x^3)\sqrt[3]{x^3 + 2}\, dx.$$

Combining the integrals then yields

$$\int (x^6 + x^3)\sqrt[3]{x^3 + 2}\, dx = \frac{1}{8} x^4 (x^3 + 2)^{4/3} + C.$$

Solution 4. (George Andrews, Pennsylvania State University)   Given that we can find an antiderivative, we expect it to have the form $p(x)\sqrt[3]{x^3 + 2}$ for some polynomial $p(x)$. Thus we look for a polynomial

$$p(x) = a_0 + a_1 x + a_2 x^2 + a_3 x^3 + \cdots$$

for which

$$\frac{d}{dx}\left(p(x)\sqrt[3]{x^3 + 2} \right) = (x^6 + x^3)\sqrt[3]{x^3 + 2}.$$

This leads to the differential equation

$$(x^3 + 2)p'(x) + x^2 p(x) = (x^6 + x^3)(x^3 + 2),$$

that is,

$$(x^3+2)(a_1+2a_2x+3a_3x^2+\cdots)+(a_0x^2+a_1x^3+a_2x^4+\cdots) = x^9+3x^6+2x^3.$$

Comparing coefficients of x^n, $n \geq 0$, and using the convention that $a_j = 0$ for $j < 0$, we find that

$$(2n + 2)a_{n+1} + (n - 1)a_{n-2} = \begin{cases} 2 & \text{if } n = 3, \\ 3 & \text{if } n = 6, \\ 1 & \text{if } n = 9, \\ 0 & \text{otherwise.} \end{cases}$$

This is equivalent to

$$a_{n+3} = -\frac{n+1}{2n+6}\,a_n + \frac{\delta_n}{2n+6}, \qquad \text{where} \quad \delta_n = \begin{cases} 2 & \text{if } n = 1, \\ 3 & \text{if } n = 4, \\ 1 & \text{if } n = 7, \\ 0 & \text{otherwise.} \end{cases}$$

In particular, for $p(x)$ to be a polynomial, we must have $a_0 = a_3 = a_6 = \cdots = 0$ and $a_2 = a_5 = a_8 = \cdots = 0$, while from $a_1 = 0$ we get $a_4 = 1/4$, $a_7 = 1/8$, $a_{10} = a_{13} = a_{16} = \cdots = 0$. Thus,

$$\int (x^6 + x^3) \sqrt[3]{x^3 + 2}\, dx = \left(\frac{1}{8}\,x^7 + \frac{1}{4}\,x^4\right) \sqrt[3]{x^3 + 2} + C.$$

Comment.   Solution 4 illustrates the *Risch algorithm*, which finds, when possible, antiderivatives of elementary functions.

Problem 13

Two microfilms are hidden in two dark squares of a chessboard; the squares are adjacent to opposite corners and symmetric to each other about the center of the board. The chessboard is a collapsible variety which folds along both center lines. Suppose the four quarters of the chessboard come apart and then are reassembled in a random manner (but so as to maintain the checkerboard pattern of light and dark squares). What is the probability that the two squares which hide the microfilms are again adjacent to opposite corners?

Solution.   The probability is $1/24$.

      When the board is reassembled so there is a light square in the lower right-hand corner, there are eight possible positions that can have a film (indicated by black dots in Figure 2). Of these eight positions, two are adjacent to a corner,

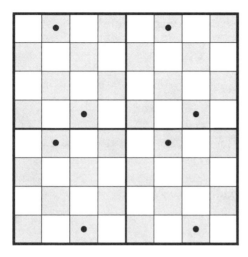

FIGURE 2

so there is a probability of $2/8 = 1/4$ that a given film will end up adjacent to a corner. If this happens, there are six possibilities remaining for the second film since the two films cannot end up in the same "quadrant" of the board. Of these six possibilities, only one is adjacent to a corner. Thus the probability of both films ending up in acceptable positions is $(1/4)(1/6) = 1/24$.

Problem 14

Let $f(x)$ be a positive, continuously differentiable function, defined for all real numbers, whose derivative is always negative. For any real number x_0, will the sequence (x_n) obtained by Newton's method $(x_{n+1} = x_n - f(x_n)/f'(x_n))$ always have limit ∞?

Solution. Yes. To see why, note that since $f'(x)$ is always negative while $f(x)$ is always positive, the sequence (x_n) is increasing. Therefore, if the sequence did not have limit ∞, it would be bounded and have a finite limit L. But then taking the limit of both sides of the equation

$$x_{n+1} = x_n - \frac{f(x_n)}{f'(x_n)}$$

(and using the fact that f is continuously differentiable) would yield

$$L = L - \frac{f(L)}{f'(L)}$$

and thus $f(L) = 0$. This is a contradiction, since f is supposed to be a positive function.

Problem 15

Find all solutions in nonnegative integers to the system of equations

$$3x^2 - 2y^2 - 4z^2 + 54 = 0, \qquad 5x^2 - 3y^2 - 7z^2 + 74 = 0.$$

Solution.   The two solutions are $(x, y, z) = (4, 7, 1)$ and $(16, 13, 11)$.

We eliminate x by subtracting 3 times the second equation from 5 times the first, obtaining $-y^2 + z^2 + 48 = 0$, or

$$48 = y^2 - z^2 = (y + z)(y - z).$$

Since $(y + z) - (y - z) = 2z$ is even, both $y + z$ and $y - z$ must be even. The possibilities are $(y + z, y - z) = (24, 2)$, $(12, 4)$, and $(8, 6)$. Solving for y and z yields $(y, z) = (13, 11)$, $(8, 4)$, and $(7, 1)$. From either original equation, we find the corresponding x values to be 16, $\sqrt{46}$, and 4, respectively. Thus, the only nonnegative integral solutions to the system of equations are $(16, 13, 11)$ and $(4, 7, 1)$.

Comment.   If we eliminate y instead, we get $x^2 - 2z^2 = 14$, a Pellian equation with infinitely many solutions, which is more difficult to solve than $y^2 - z^2 = 48$.

Problem 16

If n is a positive integer, how many real solutions are there, as a function of n, to $e^x = x^n$?

Solution.   For $n = 1$, there are no solutions; for $n = 2$, there is one solution; for odd $n \geq 3$, there are two solutions; for even $n \geq 4$, there are three solutions.

Let us first consider negative solutions. For odd n, there are none, since $x^n < 0$ when $x < 0$. So suppose n is even. Then x^n decreases for $x < 0$. Since e^x increases, $x^n - e^x$ decreases for $x < 0$. (This can also be seen by examining the derivative.) Since $x^n - e^x$ approaches ∞ as x approaches $-\infty$ and since $0^n - e^0 = -1 < 0$, there must be exactly one negative x for which $x^n - e^x = 0$.

Clearly, $x = 0$ is never a solution, so we have only positive solutions left to consider. For these there is no need to distinguish between even and odd n; in fact, we will prove the following claim:

If $a > 0$ is any real number, then $e^x = x^a$ has no positive solutions if $a < e$, one positive solution ($x = e$) if $a = e$, and two positive solutions if $a > e$.

Since 1 and 2 are the only positive integers less than e, we will be done once we prove this.

To prove the claim, consider the ratio

$$f(x) = \frac{x^a}{e^x}.$$

We are interested in the number of solutions of $f(x) = 1$. Note that $f(0) = 0$ and $\lim_{x \to \infty} f(x) = 0$. Since

$$f'(x) = \frac{x^{a-1}(a - x)}{e^x},$$

f is increasing for $0 < x < a$ and decreasing for $x > a$. Therefore, $f(x) = 1$ has two solutions if $f(a) > 1$, one if $f(a) = 1$, and none if $f(a) < 1$. The claim now easily follows from $f(a) = (a/e)^a$.

Problem 17

Does there exist a positive integer whose prime factors include at most the primes 2, 3, 5, and 7 and which ends in the digits 11? If so, find the smallest such positive integer; if not, show why none exists.

Solution. There is no such positive integer.

The last digit of a multiple of 2 is even; the last digit of a multiple of 5 is 0 or 5. Therefore, 2 and 5 cannot divide a number ending in the digits 11. Thus, we consider numbers of the form $3^m 7^n$.

A check on small numbers of this form having 1 in the units position shows that $3^4 = 81$, $3 \cdot 7 = 21$, and $7^4 = 2401$. Each of these numbers has the correct units digit, but in each case, the tens digit is even instead of 1.

Now note that integers r and $r(20k + 1)$ have the same remainder upon division by 20. Since 3^4, $3 \cdot 7$, and 7^4 have the form $20k + 1$, we conclude that if $3^m 7^n$ has the form $20k + 11$ (that is, has odd tens digit and final digit 1), then so do $3^{m-4} 7^n$, $3^{m-1} 7^{n-1}$, and $3^m 7^{n-4}$, provided the respective exponents are nonnegative.

By repeated application of this observation, we can say that if any integer $3^m 7^n$ has the form $20k + 11$, then so will one of 1, 3, 3^2, 3^3, 7, 7^2, 7^3. However, among these, only 1 has the correct last digit. Thus, $3^m 7^n$ can never have remainder 11 upon division by 20, hence cannot end in the digits 11.

Comment. This proof can be streamlined by using congruences, as follows. If $3^m 7^n = 100k + 11$ for some nonnegative integers m, n, k, then, in particular,

$$3^m 7^n \equiv 11 \pmod{20}.$$

On the other hand, since $3^4 \equiv 3 \cdot 7 \equiv 7^4 \equiv 1 \pmod{20}$, we have

$$3^m 7^n \equiv 3^{m-4} 7^n \equiv 3^{m-1} 7^{n-1} \equiv 3^m 7^{n-4} \pmod{20},$$

provided the respective exponents are nonnegative. By repeated application of these identities we find that $3^m 7^n$ is congruent modulo 20 to one of the numbers 1, 3, 3^2, 3^3, 7, 7^2, 7^3. Since none of these is 11 modulo 20, neither is $3^m 7^n$.

Problem 18

The Wohascum County Fish and Game Department issues four types of licenses, for deer, grouse, fish, and wild turkey; anyone can purchase any combination of licenses. In a recent year, (exactly) half the people who bought a grouse license also bought a turkey license. Half the people who bought a turkey license also bought a deer license. Half the people who bought a fish license also bought a grouse license, and one more than half the people who bought a fish license also bought a deer license. One third of the people who bought a deer license also bought both grouse and fish licenses. Of the people who bought deer licenses, the same number bought a grouse license as bought a fish license; a similar statement was true of buyers of turkey licenses. Anyone who bought both a grouse and a fish license also bought either a deer or a turkey license, and of these people the same number bought a deer license as bought a turkey license. Anyone who bought both a deer and a turkey license either bought both a grouse and a fish license or neither. The number of people buying a turkey license was equal to the number of people who bought some license but not a fish license. The number of people buying a grouse license was equal to the number of people buying some license but not a turkey license. The number of deer licenses sold was one more than the number of grouse licenses sold. Twelve people bought either a grouse or a deer license (or both). How many people in all bought licenses? How many licenses in all were sold?

Solution. Fourteen people bought a total of thirty-one licenses.

We let n stand for the number of people buying licenses. We let d, f, g, and t stand for the number of deer, fish, grouse, and wild turkey licenses sold, respectively. Since we will never multiply two variables, we can let dg be the number of people buying both deer and grouse licenses, $dfgt$ be the number of people buying all four licenses, etc. If we then translate the given information into equations, we get

$$\tfrac{1}{2}g = gt, \tag{1}$$

$$\tfrac{1}{2}t = dt, \tag{2}$$

$$\tfrac{1}{2}f = fg, \tag{3}$$

$$\tfrac{1}{2}f + 1 = df, \tag{4}$$

$$\tfrac{1}{3}d = dfg, \tag{5}$$

$$dg = df, \tag{6}$$

$$gt = ft, \tag{7}$$

$$fg = dfg + fgt - dfgt, \tag{8}$$

$$dfg = fgt, \tag{9}$$

$$dgt = dft = dfgt, \tag{10}$$

$$t = n - f, \tag{11}$$

$$g = n - t, \tag{12}$$

$$d = g + 1, \tag{13}$$

$$12 = g + d - dg. \tag{14}$$

We can also express n in terms of the other variables using the inclusion-exclusion principle. This yields

$$n = (d + g + f + t) - (dg + df + dt + fg + gt + ft)$$
$$+ (dfg + dgt + dft + fgt) - dfgt. \tag{15}$$

Equations (13), (12), and (11) imply that d, t, and f can be expressed in terms of n and g. Doing so and substituting the results into equations (4) and (14), we find

$$df = \tfrac{1}{2}g + 1 \qquad\qquad\qquad (16)$$

and

$$dg = 2g - 11. \qquad\qquad\qquad (17)$$

Equations (6), (16), and (17) combine to yield $g = 8$. Substituting this into our expressions for d, t, and f gives

$$d = 9, \quad t = n - 8, \quad f = 8.$$

Equations (4), (6), (2), (3), (1), and (7) yield

$$df = 5, \quad dg = 5, \quad dt = \tfrac{1}{2}n - 4, \quad fg = 4, \quad gt = 4, \quad ft = 4.$$

Equations (5), (9), (8), and (10) then yield

$$dfg = 3, \quad fgt = 3, \quad dfgt = 2, \quad dgt = 2, \quad dft = 2.$$

Finally, substitution into equation (15) implies

$$n = (17 + n) - (18 + \tfrac{1}{2}n) + 10 - 2 = 7 + \tfrac{1}{2}n.$$

We conclude that 14 people bought a total of $17 + 14 = 31$ licenses.

Problem 19

Given three lines in the plane which form a triangle, what is the set of points for which the sum of the distances to the three lines is as small as possible?

Solution.   The set consists of the vertex opposite the longest side of the triangle, assuming that side is uniquely determined. If there are exactly two such sides (of equal length), the set is the remaining (shortest) side of the triangle (including endpoints); if the triangle is equilateral, the set consists of the entire triangle and its interior.

   Let P be a point for which the sum of the three distances is minimized. If A, B, and C are the points of intersection of the three lines, then it is easy to see that P cannot be outside the triangle ABC. For if P were outside the triangle, say between the lines AB and AC, inclusive, then we could reduce the sum of the distances by moving P toward A.

   Let a, b, and c be the lengths of BC, AC, and AB, respectively. There is no loss of generality in assuming $a \le b \le c$. If K is twice the area of triangle ABC, and x, y, z are the distances from a point P to BC, AC, and AB, respectively, then we wish to minimize $x + y + z$ subject to the constraint $ax + by + cz = K$, and conceivably to other geometric constraints.

Now (x, y, z) is a nonnegative triple satisfying the constraint if and only if $(0, 0, (a/c)x + (b/c)y + z)$ is such a triple. But

$$x + y + z \geq (a/c)x + (b/c)y + z,$$

with necessary and sufficient conditions for equality given by

$$a = c \quad \text{or} \quad x = 0,$$

and

$$b = c \quad \text{or} \quad y = 0.$$

We conclude that the set P consists of C if $a \leq b < c$, the line segment BC if $a < b = c$, and the triangle ABC and its interior if $a = b = c$.

Comment. As we saw, this problem translates to a linear programming problem for which the geometry of the situation makes the solution fairly apparent. The triangle and its interior correspond to the part of the plane $ax + by + cz = K$ with $x, y, z \geq 0$. The minimum of $x + y + z$ can occur at a point, on a side of the triangle, or on the entire triangle, depending on the relative orientations of the planes $ax + by + cz = K$ and $x + y + z = K/c$.

Problem 20

Suppose the plane $x + 2y + 3z = 0$ is a perfectly reflecting mirror. Suppose a ray of light shines down the positive x-axis and reflects off the mirror. Find the direction of the reflected ray.

Answer. The ray has the direction of the unit vector $\frac{1}{7}(-6, 2, 3)$.

Solution 1. Let **u** be the unit vector in the direction of the reflected ray. The vector $(1, 2, 3)$ is normal to the plane, so by vector addition and geometry, $(1, 0, 0) + \mathbf{u}$ is twice the projection of the vector $(1, 0, 0)$ onto $(1, 2, 3)$. Using the dot product formula for projection, we obtain

$$(1, 0, 0) + \mathbf{u} = 2 \frac{(1, 0, 0) \cdot (1, 2, 3)}{(1, 2, 3) \cdot (1, 2, 3)} (1, 2, 3) = \frac{1}{7}(1, 2, 3),$$

or $\mathbf{u} = \frac{1}{7}(-6, 2, 3)$.

Solution 2.   (Murray Klamkin, University of Alberta)   Let (u, v, w) denote the reflection (mirror image) of the point $(1, 0, 0)$ in the plane $x + 2y + 3z = 0$. Since the incoming ray is directed from $(1, 0, 0)$ to $(0, 0, 0)$, the reflected ray is in the direction from (u, v, w) to $(0, 0, 0)$, that is, in the direction of the vector $(-u, -v, -w)$.

To find (u, v, w), note that the vector from $(1, 0, 0)$ to (u, v, w) is normal to the plane, hence

$$(u - 1, v, w) = \lambda(1, 2, 3) \qquad \text{for some } \lambda.$$

This yields $u = \lambda + 1$, $v = 2\lambda$, $w = 3\lambda$. Also, (u, v, w) has the same distance to the origin as $(1, 0, 0)$, so $u^2 + v^2 + w^2 = 1$. Substituting for u, v, w and solving for λ, we find $\lambda = -\frac{1}{7}$ (since $\lambda \neq 0$). Therefore, $(u, v, w) = (\frac{6}{7}, -\frac{2}{7}, -\frac{3}{7})$, and the reflected ray has the direction of $\frac{1}{7}(-6, 2, 3)$.

Problem 21

Find the set of all solutions to

$$x^{y/z} = y^{z/x} = z^{x/y},$$

with x, y, and z positive real numbers.

Solution.   The only solutions are $x = y = z$, with z equal to an arbitrary positive number. These are obviously solutions; we show there are no others.

For positive numbers a and b, $a^b < 1$ if and only if $a < 1$. Thus, if one of x, y, z is less than 1, all three are. If (x, y, z) is a solution to the system of equations, then it is easy to verify that $(1/x, 1/z, 1/y)$ is also a solution. Therefore, we may assume $x, y, z \geq 1$. The symmetry of the equations allows us to assume $x \leq z$ and $y \leq z$ (but then we may not assume any relationship between x and y). We then have

$$x \geq x^{y/z} = y^{z/x} \geq y.$$

This in turn implies

$$x \geq x^{y/z} = z^{x/y} \geq z,$$

hence $x = z$. This makes the original equations

$$x^{y/x} = y = x^{x/y}.$$

If $x = 1$, then $y = 1$. If $x > 1$, then $y/x = x/y$, hence $x = y$. In either case, $x = y = z$, as asserted.

Problem 22

Find all perfect squares whose base 9 representation consists only of ones.

Solution.   The only such square is 1.

We are asked to find nonnegative integers x and n such that

$$x^2 = 1 + 9 + 9^2 + \cdots + 9^n.$$

Summing the finite geometric series on the right and multiplying by 8, we arrive at the equation

$$8x^2 = 9^{n+1} - 1 = (3^{n+1} + 1)(3^{n+1} - 1).$$

Because the greatest common divisor of $3^{n+1} + 1$ and $3^{n+1} - 1$ is 2, there must exist positive integers c and d such that either

(i) $3^{n+1} + 1 = 4c^2$ and $3^{n+1} - 1 = 2d^2$

or

(ii) $3^{n+1} + 1 = 2c^2$ and $3^{n+1} - 1 = 4d^2$.

In case (i), $3^{n+1} = 4c^2 - 1 = (2c + 1)(2c - 1)$. Since $2c + 1$ and $2c - 1$ are relatively prime, we must have $2c + 1 = 3$, $2c - 1 = 1$, and it follows that $n = 0$ and $x = 1$.

In case (ii), n must be odd, since $3^{n+1} - 1$ is divisible by 4. But then

$$1 = 3^{n+1} - 4d^2 = \left(3^{(n+1)/2} + 2d\right)\left(3^{(n+1)/2} - 2d\right),$$

which is impossible for $d > 0$.

Problem 23

In a recent snowmobile race, conditions were such that no more than two competitors could be abreast of each other anywhere along the trail. Nevertheless, there was frequent passing. The following is an excerpt from an article about the race. "After a few miles A pulled ahead of the pack, closely followed by

B and C in that order, and thereafter these three did not relinquish the top three positions in the field. However, the lead subsequently changed hands nine times among these three; meanwhile, on eight different occasions the vehicles that were running second and third at the time changed places. ... At the end of the race, C complained that B had driven recklessly just before the finish line to keep C, who was immediately behind B at the finish, from passing. ..." Can this article be accurate? If so, can you deduce who won the race? If the article cannot be accurate, why not?

Solution. The article cannot be accurate.

At the point where A, B, and C pulled ahead of the pack, the order of the three was (A, B, C): A first, B second, and C third. At the end of the race, if the article is correct, the order must have been (A, B, C) or (B, C, A). Once the three pulled ahead of the pack, there were $9 + 8 = 17$ place changes among A, B, and C. If we let the set S consist of the three orders (A, B, C), (B, C, A), and (C, A, B), and T consist of the other three possible orders, (A, C, B), (B, A, C), and (C, B, A), then a single interchange of two of the snowmobiles will take an order in S to an order in T, and vice versa. Thus, to get from (A, B, C) to either itself or to (B, C, A) requires an even number of interchanges, rather than 17.

Comment. This solution proves a specific case of the following well-known theorem on permutations: A given permutation can never be obtained both by an even number of transpositions (interchanges of two elements) and by an odd number of transpositions.

Problem 24

Show how to construct the midpoint of any given segment using only a straight-edge (but no compasses) and a "trisector," a device with which any straight line segment can be divided into three equal parts.

Solution 1. (See Figure 3.) Let AB denote the line segment we wish to bisect and let C be an arbitrary point off the line AB. Trisect AC at D and E, D being the closer point to C. Then E is the midpoint of AD, and BE is the median of the triangle ABD. Trisect the segment BE, letting F be the point two–thirds of the way to E. It is well known that F is the intersection of the medians of ABD. Hence, DF intersects AB at its midpoint.

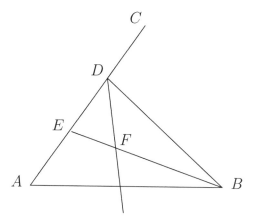

FIGURE 3

Solution 2. (See Figure 4.) Let AB be the line segment whose midpoint is to be constructed. Choose an arbitrary point C off the line AB; draw the line segments AC and BC. Trisect AC and BC; let P and Q be the trisection points closest to C on AC and BC, respectively. Now draw the line segments AQ and BP; let R be their point of intersection. The line CR will then intersect the segment AB at the desired midpoint M. This follows immediately from Ceva's theorem.

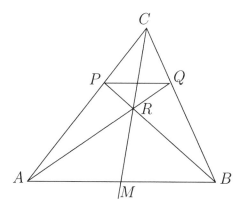

FIGURE 4

Comment. To prove the validity of the second construction without recourse to Ceva's theorem, one can use the following general result: Let A, B, and C be noncollinear. Let P lie on the line segment AC and Q lie on the line segment BC such that PQ is parallel to AB. Let R be the intersection of AQ and BP. Then R lies on the median from C to AB. To prove this, let X on AC and Y on BC be on the line parallel to AB through R. Then the triangles PXR and PAB are similar, as are the triangles QRY and QAB. The ratios XR/AB and RY/AB are both equal to the ratio of the distance between the lines PQ and XY to the distance between the lines PQ and AB, hence are equal. Thus, $XR = RY$. If M is the intersection of AB and CR, the similarity of triangles CXR and CAM and of CRY and CMB imply that M is the midpoint of AB, proving the result.

Problem 25

Consider a 12×12 chessboard (consisting of 144 1×1 squares). If one removes 3 corners, can the remainder be covered by 47 1×3 tiles?

Idea. This is reminiscent of the old "chessnut": If opposite corners of an ordinary 8×8 chessboard are removed, the modified board cannot be covered with 31 1×2 tiles. The reason is that if the chessboard is colored in the usual alternating manner, opposite corners will have the same color. Thus, the modified board will consist of 30 squares of one color and 32 of the other. Since any tile in a tiling will cover one white square and one black square, no tiling is possible. We look for a similar coloring idea: What about three colors?

Solution. No. To see why, rotate the 12×12 chessboard so the non-deleted corner square is in the upper right corner, and color the board as shown in Figure 5.

The board has 48 squares of each color, labeled here as 1, 2, and 3. We have deleted one square of color 1 and two of color 3, so 47 of the remaining squares are labeled 1, 48 are labeled 2, and 46 are labeled 3. Thus the board cannot be covered by 1×3 tiles, because for any such tiling there would be an equal number of squares of each color.

Comment. Our coloring scheme would not work if the only remaining corner were the lower right corner.

	1	2	3	1	2	3	1	2	3	1	2	
2	3	1	2	3	1	2	3	1	2	3	1	
1	2	3	1	2	3	1	2	3	1	2	3	
3	1	2	3	1	2	3	1	2	3	1	2	
2	3	1	2	3	1	2	3	1	2	3	1	
1	2	3	1	2	3	1	2	3	1	2	3	
3	1	2	3	1	2	3	1	2	3	1	2	
2	3	1	2	3	1	2	3	1	2	3	1	
1	2	3	1	2	3	1	2	3	1	2	3	
3	1	2	3	1	2	3	1	2	3	1	2	
2	3	1	2	3	1	2	3	1	2	3	1	
	2	3	1	2	3	1	2	3	1	2		

FIGURE 5

Problem 26

Is there a function f, differentiable for all real x, such that

$$|f(x)| < 2 \qquad \text{and} \qquad f(x)f'(x) \geq \sin x \, ?$$

Idea. Observe that $2f(x)f'(x)$ is the derivative of $(f(x))^2$, so we should be thinking "integration."

Solution. There is no such function. To see this, note that for $x > 0$, the second condition implies

$$\big(f(x)\big)^2 - \big(f(0)\big)^2 = \int_0^x 2f(t)f'(t)\,dt \geq \int_0^x 2\sin t\,dt = 2 - 2\cos x.$$

Thus, for $x > 0$,

$$\big(f(x)\big)^2 \geq 2 - 2\cos x + \big(f(0)\big)^2 \geq 2 - 2\cos x.$$

Therefore, in particular, $(f(\pi))^2 \geq 4$, so $|f(\pi)| \geq 2$, contradicting the first condition.

Comment. If the condition $|f(x)| < 2$ were replaced with $|f(x)| \leq 2$, then $f(x) = 2\sin(x/2)$ would satisfy both conditions.

Problem 27

Let N be the largest possible number that can be obtained by combining the digits 1, 2, 3 and 4 using the operations addition, multiplication, and exponentiation, if the digits can be used only once. Operations can be used repeatedly, parentheses can be used, and digits can be juxtaposed (put next to each other). For instance, 12^{34}, $1 + (2 \times 3 \times 4)$, and $2^{31 \times 4}$ are all candidates, but none of these numbers is actually as large as possible. Find N.

Solution. $N = 2^{(3^{4^1})}$; by a standard convention, the parentheses are actually unnecessary. We find that $N \approx 10^{1.1 \times 10^{19}}$, which means that N has between 10^{19} and 10^{20} digits!

We first show that although addition and multiplication are allowed, these operations are not needed in forming N. In fact, for all $a, b \geq 2$, we have $a+b \leq ab \leq a^b$, so any addition or multiplication involving only (integer) terms or factors greater than 1 can be replaced by an exponentiation. Meanwhile, multiplication by 1 is useless, and the result of an addition of the form $a + 1$ will certainly be less than the result of juxtaposing 1 with one of the digits within the expression for a.

It follows that N can be formed by juxtaposition and exponentiation only. Clearly at least one exponentiation is needed (else $N \leq 4321$), and so N must be of the form a^b. Suppose the base, a, is formed by exponentiation, say $a = c^d$. Note that $b, d \geq 2$. Then $a^b = c^{db} \leq c^{(d^b)}$. Thus we may assume that a is not formed by exponentiation, but consists of one or more juxtaposed digits. If a has three digits, then b can only have one, and $a^b < 1000^4 = 10^{12}$; if a has two digits, then b is formed from the other two, so $b \leq 3^4 = 81$, and $a^b < 100^{81} = 10^{162}$. In both these cases, a^b is much less than $2^{3^{41}}$. Therefore, a must consist of one digit, and so $a = 2, 3,$ or 4.

Similar arguments to those above show that in forming b from the three remaining digits, 1 will be juxtaposed with one of the other two digits and the result will be used in an exponentiation. It is also clear that 1 will be the second digit in the juxtaposition. We now have the following possibilities: If $a = 2$, then $b = 3^{41}, 4^{31}, 41^3,$ or 31^4; if $a = 3$, then $b = 2^{41}, 4^{21}, 21^4,$ or 41^2; if $a = 4$, then $b = 3^{21}, 2^{31}, 21^3,$ or 31^2. Using the inequality $3^b < 4^b = 2^{2b}$, it is easy to shorten this list to $a = 2$ and b either 3^{41} or 4^{31}. We now use the inequality

$$4^{31} = 4^{24} \cdot 4^7 = 8^{16} \cdot 16394 < 9^{16} \cdot 19683 = 3^{41},$$

and we are done.

Problem 28

Babe Ruth's batting performance in the 1921 baseball season is often considered the best in the history of the game. In home games, his batting average was .404; in away games it was .354. Furthermore, his slugging percentage at home was a whopping .929, while in away games it was .772. This was based on a season total of 44 doubles, 16 triples, and 59 home runs. He had 30 more at bats in away games than in home games. What were his overall batting average and his slugging percentage for the year?

Solution. Babe Ruth's batting average was .378; his slugging percentage was .846. To find these powerful numbers, let a_1, h_1, and t_1 be the number of at bats, hits, and total bases (hits $+$ doubles $+ 2 \times$ triples $+ 3 \times$ home runs), respectively, in home games, and let a_2, h_2 and t_2 be the corresponding numbers for away games.

We first observe that

$$a_2 = a_1 + 30 \quad \text{and} \quad (t_1 + t_2) - (h_1 + h_2) = 44 + 2 \cdot 16 + 3 \cdot 59 = 253.$$

The given batting averages tell us that

$$.4035\, a_1 \leq h_1 \leq .4045\, a_1 \quad \text{and} \quad .3535\,(a_1 + 30) \leq h_2 \leq .3545\,(a_1 + 30).$$

Adding these inequalities, we get

$$.757\, a_1 + 10.605 \ \leq h_1 + h_2 \leq .759\, a_1 + 10.635. \tag{$*$}$$

Similarly, from the given slugging percentages we find

$$.9285\, a_1 \leq t_1 \leq .9295\, a_1 \quad \text{and} \quad .7715\,(a_1 + 30) \leq t_2 \leq .7725\,(a_1 + 30),$$

hence

$$1.7\, a_1 + 23.145 \ \leq \ t_1 + t_2 \ \leq 1.702\, a_1 + 23.175.$$

Substituting $h_1 + h_2 + 253$ for $t_1 + t_2$, we obtain

$$1.7\, a_1 - 229.855 \ \leq h_1 + h_2 \ \leq 1.702\, a_1 - 229.825. \tag{$**$}$$

Now we can combine $(*)$ and $(**)$ to get, on the one hand,

$$1.7\, a_1 - 229.855 \ \leq \ .759\, a_1 + 10.635,$$

which yields $.941\, a_1 \leq 240.49$, or $a_1 \leq 255.56\ldots$. Since a_1 is an integer, $a_1 \leq 255$. On the other hand,

$$.757\, a_1 + 10.605 \ \leq \ 1.702\, a_1 - 229.825,$$

from which we conclude $254.42\ldots \leq a_1$, hence $255 \leq a_1$. Therefore, $a_1 = 255$. With this, (∗) easily yields $h_1 + h_2 = 204$, and there is no further difficulty in computing the overall batting average $(h_1 + h_2)/(a_1 + a_2)$ and slugging percentage $(h_1 + h_2 + 253)/(a_1 + a_2)$ to be .378 and .846, respectively.

Comment. From a numerical point of view, Babe Ruth's statistics as given are not very delicate. If we set $h_1 = .404\,a_1$, $h_2 = .354\,(a_1 + 30)$, $t_1 = .929\,a_1$, $t_2 = .772\,(a_1 + 30)$, and $t_1 + t_2 - h_1 - h_2 = 253$, and solve this linear system, we obtain $h_1 \approx 103.02$, $h_2 \approx 100.89$, $t_1 \approx 236.89$, $t_2 \approx 220.02$, $a_1 \approx 254.99$. It is then easy to guess the correct answer. However, to justify this, analysis of the round-off error is required. To give a simple example, the system $x + y = 2$, $x + 1.001y = 2$ has $x = 2$, $y = 0$ as its solution, whereas the system $x + y = 2$, $x + 1.001y = 2.001$ has solution $x = y = 1$. In our case, the coefficients have been rounded, as well. For more information, see a text on numerical analysis or on applied linear algebra.

Problem 29

Let C be a circle with center O, and Q a point inside C different from O. Where should a point P be located on the circumference of C to maximize $\angle OPQ$?

Answer. The angle $\angle OPQ$ is a maximum when $\angle OQP$ is a right angle.

Solution 1. (Jason Colwell, age 13, Edmonton, Canada.)   Think of P as fixed, while Q varies along the circle about O with radius OQ. It is then clear

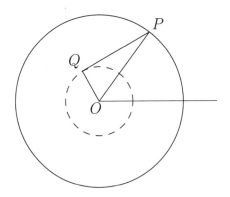

FIGURE 6

that $\angle OPQ$ is a maximum when PQ is tangent to the small circle, that is, when $\angle OQP = 90°$.

Solution 2. Let $\theta = \angle OPQ$ and $\varphi = \angle OQP$, as shown in Figure 7.

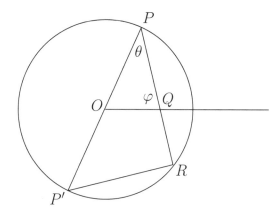

FIGURE 7

By the law of sines applied to $\triangle OQP$,

$$\frac{OP}{\sin \varphi} = \frac{OQ}{\sin \theta} \qquad \text{or} \qquad \sin \theta = \frac{OQ}{OP} \sin \varphi \le \frac{OQ}{OP}$$

with equality if and only if $\varphi = 90°$. Since $\frac{OQ}{OP}$ is constant, $\sin \theta$, and hence also θ, is maximized when $\angle OQP$ is a right angle.

Solution 3. Extend PO and PQ to intersect the circle C at P' and R, respectively. For any point P, PP' is a diameter of C, and therefore $\angle PRP'$ is a right angle. Now θ will be maximized when $\cos \theta$ is minimized; $\cos \theta$ is minimized when chord PR is minimized. Finally, PR is minimized when it is perpendicular to OQ, since any other chord through Q passes closer to O, and this completes the proof.

Problem 30

Find all real solutions of the equation $\sin(\cos x) = \cos(\sin x)$.

Solution. There are no real solutions.

Note that each side of the equation has period 2π. Therefore, we can assume $-\pi < x \leq \pi$. If x is a solution, then

$$\sin(\cos(-x)) = \sin(\cos x) = \cos(\sin x) = \cos(-\sin x) = \cos(\sin(-x)),$$

hence $-x$ is also a solution. Thus, it suffices to show there are no solutions with $0 \leq x \leq \pi$. On this interval, $0 \leq \sin x \leq 1 < \pi/2$, so we see that $\cos(\sin x) > 0$. In order to have $\sin(\cos x) > 0$, we must have $0 \leq x < \pi/2$. Since the function $x - \sin x$ is increasing and is 0 when $x = 0$, we have $\sin x < x$ for $x > 0$; in particular, $\sin(\cos x) < \cos x$. On the other hand, since the cosine function decreases on $[0, \pi/2)$, we have $\cos x \leq \cos(\sin x)$, so $\sin(\cos x) < \cos(\sin x)$, and we are done.

Problem 31

For a natural number $n \geq 2$, let $0 < x_1 \leq x_2 \leq \cdots \leq x_n$ be real numbers whose sum is 1. If $x_n \leq 2/3$, prove that there is some k, $1 \leq k \leq n$, for which $1/3 \leq \sum_{j=1}^{k} x_j < 2/3$.

Solution. (Allen Schwenk, Western Michigan University)   We distinguish two cases.

Case 1. $x_n > 1/3$. Then $1/3 \leq 1 - x_n < 2/3$, and since $1 - x_n = \sum_{j=1}^{n-1} x_j$, we can take $k = n - 1$.

Case 2. $x_n \leq 1/3$. Define $s_k = \sum_{j=1}^{k} x_j$. If $1/3 \leq s_k < 2/3$ for any k, we are done. If not, since $s_0 = 0$ and $s_n = 1$, we may select the smallest k with $s_k \geq 2/3$. Then presumably $s_{k-1} < 1/3$. But this forces

$$x_k = s_k - s_{k-1} > 1/3 \geq x_n,$$

a contradiction.

Problem 32

Is there a cubic curve $y = ax^3 + bx^2 + cx + d$, $a \neq 0$, for which the tangent lines at two distinct points coincide?

Answer. No. For each of the solutions below, we assume (x_1, y_1) and (x_2, y_2), $x_1 < x_2$, are two such points on the cubic curve.

Solution 1. By the Mean Value Theorem, there exists x_3, with $x_1 < x_3 < x_2$, such that $(y_2 - y_1)/(x_2 - x_1) = y'(x_3)$. Since the tangent lines coincide, we have $y'(x_1) = y'(x_2) = y'(x_3) = M$, say. But then $3ax^2 + 2bx + c - M$ is a quadratic polynomial with 3 distinct roots, x_1, x_2, and x_3, and this is impossible.

Solution 2. If the tangent line through (x_1, y_1) and (x_2, y_2) is $y = Mx + B$, then x_1 and x_2 are multiple roots of the equation

$$ax^3 + bx^2 + (c - M)x + (d - B) = 0.$$

This follows from the fact that r is a multiple root of a polynomial $p(x)$ if and only if $p(r) = p'(r) = 0$. However, a cubic polynomial cannot be divisible by the fourth degree polynomial $(x - x_1)^2(x - x_2)^2$, so we have a contradiction.

(Recall that r is a *multiple root* of a polynomial $p(x)$ if $p(x)$ is divisible by $(x - r)^2$.)

Problem 33

A group of friends had n ($n \geq 3$) digital watches between them, each perfectly accurate in the sense that a second was the same according to each watch. The time indicated shifted from one watch to the next, but in such a way that any two watches showed the same time in minutes for part of each minute. Find the largest number x such that at least one pair of watches necessarily showed the same time in minutes for more than the fraction x of each minute. (This is part (b) of the original problem; part (a) will follow by taking $n = 3$.)

Solution. The largest number with the stated property is $x = (n-2)/(n-1)$; in particular, for $n = 3$ we have $x = 1/2$.

Order the watches so that the first watch is running ahead of (or possibly with) all of the others, the second ahead is of all but the first, ..., and the nth is behind all the rest. Let t_j denote the time in minutes that watch j runs behind watch 1. Our ordering and the fact that all pairs of watches show the same time for some part of every minute implies

$$0 = t_1 \leq t_2 \leq \cdots \leq t_n < 1.$$

Since

$$t_n = \sum_{j=1}^{n-1}(t_{j+1} - t_j) < 1,$$

we must have

$$0 \le t_{j+1} - t_j < \frac{1}{n-1}$$

for some j. For that j, watches j and $j+1$ agree for more than the fraction

$$1 - \frac{1}{n-1} = \frac{n-2}{n-1}$$

of each minute.

Conversely, if

$$t_j = \frac{j-1}{n-1}t_n,$$

then no pair of watches agree for more than $\frac{n-2}{n-1}t_n$. Since t_n can be arbitrarily close to 1, $\frac{n-2}{n-1}$ is the best possible result.

Problem 34

Let C be a circle with center O, and Q a point inside C different from O. Show that the area enclosed by the locus of the centroid of triangle OPQ as P moves about the circumference of C is independent of Q.

Solution. We will show that the locus of the centroid of triangle OQP is a circle whose radius is one-third that of the circle C. Thus, regardless of the location of Q inside C, the area enclosed by the locus of the centroid is one-ninth that of C.

Introduce rectangular coordinates so that O is the origin, C has equation $x^2 + y^2 = r^2$, and $Q = (q, 0)$, $0 < q < r$. Let $P = (x, y)$ be an arbitrary point on C. The midpoint D of PQ has coordinates $D = ((x+q)/2, y/2)$, and the centroid, G, lies two-thirds of the distance along the median line OD, so its coordinates are $G = ((x+q)/3, y/3)$. We have

$$\left(\frac{x+q}{3} - \frac{q}{3}\right)^2 + \left(\frac{y}{3}\right)^2 = \left(\frac{x}{3}\right)^2 + \left(\frac{y}{3}\right)^2 = \left(\frac{r}{3}\right)^2,$$

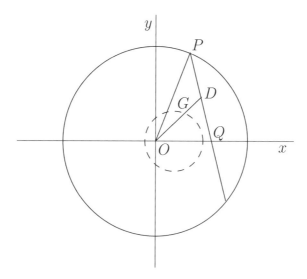

FIGURE 8

and we see that the locus of the centroid is the circle with center $(q/3, 0)$ and radius $r/3$.

Comment. We can streamline this solution by using complex numbers, as follows. Let O be the origin in the complex plane, let r be the radius of the circle, and let Q correspond to the complex number α.

Any point P on the circle corresponds to $re^{i\theta}$ for some θ; the centroid of triangle OPQ is then given by the complex number

$$z = \frac{0 + re^{i\theta} + \alpha}{3}.$$

Thus, $|z - \alpha/3| = r/3$. This is the equation of a circle of radius $r/3$, centered at $\alpha/3$.

Problem 35

Describe the set of points (x, y) in the plane for which

$$\sin(x + y) = \sin x + \sin y.$$

Answer. The set of solutions consists of three infinite families of parallel lines: $x = 2k\pi$, $y = 2k\pi$, $x + y = 2k\pi$, with k an integer.

Solution 1. The relation is certainly true if $x = 2k\pi$ (with k an integer) or if $y = 2k\pi$, so the graph contains the infinitely many horizontal lines $y = 2k\pi$ and vertical lines $x = 2k\pi$.

Now assume that x and y are not (integer) multiples of 2π. We can rewrite the relation as

$$\sin x \cos y + \cos x \sin y = \sin x + \sin y$$

or

$$\sin x(\cos y - 1) = \sin y(1 - \cos x).$$

Since x and y are not multiples of 2π, $1 - \cos x \neq 0$ and $\cos y - 1 \neq 0$, so we can divide by these factors, obtaining

$$\frac{\sin x}{1 - \cos x} = \frac{\sin y}{\cos y - 1} = \frac{-\sin(-y)}{\cos(-y) - 1} = \frac{\sin(-y)}{1 - \cos(-y)}.$$

Let $f(x) = \sin x/(1 - \cos x)$. We have seen that (except for the special cases $x = 2k\pi$, $y = 2k\pi$) the relation is true if and only if $f(x) = f(-y)$.

Now

$$f(x) = \frac{\sin x}{1 - \cos x} = \frac{2\sin(x/2)\,\cos(x/2)}{2\sin^2(x/2)} = \cot(x/2),$$

and $\cot(x/2)$ has period 2π and decreases from ∞ to $-\infty$ on each interval between successive integer multiples of 2π. So on each such interval, $f(x)$ takes on each value only once. Hence, $f(x) = f(-y)$ if and only if $x = -y + 2k\pi$ for some integer k; that is, when $x + y = 2k\pi$ for some integer k. This completes the proof.

Solution 2. (William Firey, Oregon State University)   Use trigonometric identities to rewrite the equation as

$$2\sin\left(\frac{x+y}{2}\right)\cos\left(\frac{x+y}{2}\right) = 2\sin\left(\frac{x+y}{2}\right)\cos\left(\frac{x-y}{2}\right).$$

This implies that either

$$\sin\left(\frac{x+y}{2}\right) = 0,$$

resulting in $x + y = 2k\pi$, k an integer, or

$$\cos\left(\frac{x+y}{2}\right) = \cos\left(\frac{x-y}{2}\right),$$

$$\cos(x/2)\cos(y/2) - \sin(x/2)\sin(y/2) = \cos(x/2)\cos(y/2) + \sin(x/2)\sin(y/2),$$

$$\sin(x/2)\sin(y/2) = 0,$$

which implies that $x = 2k\pi$ or $y = 2k\pi$, k an integer.

Problem 36

Show that every 2×2 matrix of determinant 1 is the product of *three* elementary matrices.

Idea. The product of upper triangular matrices is upper triangular; also, the product of lower triangular matrices is lower triangular. Thus, a factorization into a product of elementary matrices must generally involve matrices of type $\left(\begin{smallmatrix} 1 & x \\ 0 & 1 \end{smallmatrix}\right)$ and of type $\left(\begin{smallmatrix} 1 & 0 \\ x & 1 \end{smallmatrix}\right)$. On the other hand, if a factorization only had one matrix of type $\left(\begin{smallmatrix} y & 0 \\ 0 & 1 \end{smallmatrix}\right)$ or $\left(\begin{smallmatrix} 1 & 0 \\ 0 & y \end{smallmatrix}\right)$, the determinant would be y, so it is probably not worth including any such matrices.

Solution. Let $\left(\begin{smallmatrix} a & b \\ c & d \end{smallmatrix}\right)$ be a matrix of determinant 1. We first try to write

$$\begin{pmatrix} a & b \\ c & d \end{pmatrix} = \begin{pmatrix} 1 & x \\ 0 & 1 \end{pmatrix}\begin{pmatrix} 1 & 0 \\ y & 1 \end{pmatrix}\begin{pmatrix} 1 & z \\ 0 & 1 \end{pmatrix}.$$

By straightforward computation, this is equivalent to

$$a = 1 + xy, \quad b = z(1 + xy) + x, \quad c = y, \quad d = yz + 1. \qquad (*)$$

If $c \neq 0$, then we can solve $(*)$ for y, x, z in that order using all but the second equation; the result is $y = c$, $x = (a-1)/c$, $z = (d-1)/c$. For these choices of x, y, z the second equation in $(*)$ is satisfied as well, since

$$z(1 + xy) + x = c^{-1}(d-1)(1 + a - 1) + c^{-1}(a-1) = c^{-1}(ad - 1) = b,$$

where we have used $\det\left(\begin{smallmatrix} a & b \\ c & d \end{smallmatrix}\right) = 1$. So we are done unless $c = 0$. If $c = 0$, the matrix will be of the form

$$\begin{pmatrix} a & b \\ 0 & a^{-1} \end{pmatrix} = \begin{pmatrix} a & 0 \\ 0 & 1 \end{pmatrix}\begin{pmatrix} 1 & b \\ 0 & 1 \end{pmatrix}\begin{pmatrix} 1 & 0 \\ 0 & a^{-1} \end{pmatrix}.$$

Problem 37

Let $ABCD$ be a convex quadrilateral. Find a necessary and sufficient condition for a point P to exist inside $ABCD$ such that the four triangles ABP, BCP, CDP, DAP all have the same area.

Answer. A necessary and sufficient condition is for one of the diagonals AC and BD to bisect the other.

Solution 1. To prove sufficiency, let Q be the intersection of AC and BD. Without loss of generality, we may assume that $AQ = CQ$. Let P be the midpoint of BD. Furthermore, let R be the foot of the perpendicular from A to BD and let S be the foot of the perpendicular from C to BD.

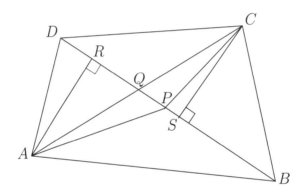

FIGURE 9

Since $BP = DP$ and since AR is the altitude for triangles ABP and DAP, these triangles have equal area. Similarly, the areas of triangles BCP and CDP are equal. Triangles ARQ and CSQ are congruent; therefore, $AR = CS$ and, hence, the areas of triangles ABP and BCP are equal. We conclude that the four triangles ABP, BCP, CDP, and DAP all have the same area.

To prove the necessity, assume that P is a point in the interior of $ABCD$ for which triangles ABP, BCP, CDP, and DAP have the same area (see Figure 10). Since triangles ABP and BCP have the same area, vertices A and C are equidistant from line BP. Because the quadrilateral $ABCD$ is convex, A and C lie on opposite sides of the line BP. An argument using congruent triangles (similar to the one in the sufficiency proof above) shows that the line

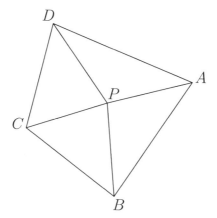

FIGURE 10

BP must actually pass through the midpoint of the line segment AC. Similarly, the line DP must also pass through the midpoint of AC. If B, P, and D are not collinear, then P must be the midpoint of AC. But then, since BCP and CDP have the same area, B and D are equidistant from the line AC and thus AC bisects the segment BD. If B, P, and D are collinear, then P must be the midpoint of BD. In this case, A and C are equidistant from the line BD, hence BD bisects the segment AC, completing the proof.

Solution 2. We prove the sufficiency as above. To prove the necessity, let θ_1, θ_2, θ_3, and θ_4 be the measures of the angles APB, BPC, CPD, and DPA, respectively (see Figure 11). The area of an arbitrary triangle is half the product of two sides and the sine of their included angle. The product of the areas of triangles APB and CPD equals the product of the areas of BPC and DPA. Thus,

$$\sin\theta_1 \sin\theta_3 = \sin\theta_2 \sin\theta_4,$$

or, from the formula for the cosine of a sum,

$$\cos(\theta_1 - \theta_3) - \cos(\theta_1 + \theta_3) = \cos(\theta_2 - \theta_4) - \cos(\theta_2 + \theta_4).$$

Since $\theta_1 + \theta_2 + \theta_3 + \theta_4 = 2\pi$, this implies

$$\cos(\theta_1 - \theta_3) = \cos(\theta_2 - \theta_4).$$

There is no loss of generality in assuming

$$\theta_3 \leq \theta_1 < \pi \quad \text{and} \quad \theta_4 \leq \theta_2 < \pi$$

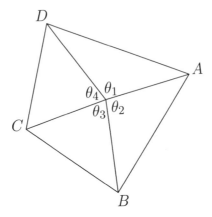

(if necessary, rename the vertices and/or reflect the quadrilateral). Under these conditions, $\theta_1 - \theta_3 = \theta_2 - \theta_4$, hence $\theta_1 + \theta_4 = \theta_2 + \theta_3$, from which we conclude that B, P, and D are collinear. As in the first solution, this implies BD bisects AC.

Problem 38

Suppose that n dancers, n even, are arranged in a circle so that partners are directly opposite each other. During the dance, two dancers who are next to each other change places while all others stay in the same place; this is repeated with different pairs of adjacent dancers until, in the ending position, the two dancers in each couple are once again opposite each other, but in the opposite of the starting position (that is, every dancer is halfway around the circle from her/his original position). What is the least number of interchanges (of two adjacent dancers) necessary to do this?

Solution. The least number of interchanges required is $n^2/4$.

   If the dancers are numbered $1, 2, \ldots, n$ as we move clockwise around the circle, first begin by interchanging dancer number $n/2$ with dancer $n/2+1$, then with dancer $n/2 + 2, \ldots$, and finally with dancer n. We have made $n/2$ interchanges and dancer $n/2$ is in the correct position. Next, interchange dancer $n/2 - 1$ with dancers $n/2 + 1, n/2 + 2, \ldots, n$. Continuing this process until dancer 1 has switched with dancers $n/2+1, n/2+2, \ldots, n$, we arrive at the desired position, having made $(n/2)^2 = n^2/4$ interchanges. On the other hand, each dancer starts $n/2$ places away from his or her final position, and thus must participate in at least $n/2$ interchanges. Each interchange is between two dancers; therefore, at least $(n \cdot n/2)/2 = n^2/4$ interchanges are required, completing the proof.

Comments. Another systematic way to move each dancer halfway around the circle using $n^2/4$ interchanges is as follows. As a first step, interchange the dancers in positions 1 and 2, those in positions 3 and 4, etc., for $n/2$ interchanges in all. In the second step, interchange the dancers now in positions 2 and 3, those in positions 4 and 5, etc. The third step is like the first, the fourth is like the second, etc. After $n/2$ steps, or $n^2/4$ interchanges, all dancers will have shifted either $n/2$ places forward, or $n/2$ places backward, so they are halfway around the circle from their original positions.

To the best of our knowledge, $n^2/4$ is the largest number of interchanges required to arrive at an arrangement of dancers from any other arrangement.

Problem 39

Let L be a line in the plane; let A and B be points on L which are a distance 2 apart. If C is any point in the plane, there may or may not (depending on C) be a point X on the line L for which the distance from X to C is equal to the average of the distances from X to A and B. Give a precise description of the set of all points C in the plane for which there is no such point X on the line.

Answer. The set of all points C in the plane for which there is no such point X consists of those points on the perpendicular bisector of AB whose distance from AB is greater than 1 (see Figure 12).

Solution 1. Let M denote the midpoint of AB. First note that if X is a point on the line L, the average of the distances from X to A and B is the distance from X to M if X is outside the line segment AB, and 1 if X is on the line segment AB. Therefore, for the distance from X to C to equal this average, we must have X at the same distance to M and C if X is outside AB, while we must have X at distance 1 to C if X is on AB.

It is easy to show that the set of points C for which there is a point X on AB with distance 1 to C is the closed set Γ shown in Figure 13. (The boundary of Γ consists of two line segments parallel to AB on either side of AB with distance

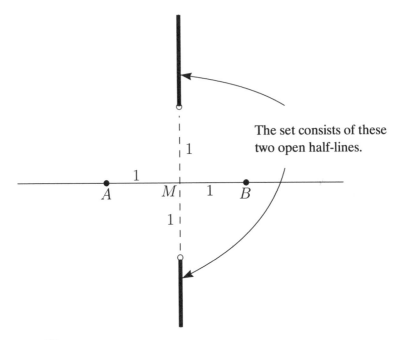

The set consists of these two open half-lines.

FIGURE 12

1 to AB and the same length as AB, and two semicircles. The semicircles are centered at A and at B; they have radius 1, and the diameter connecting the end points of each semicircle is perpendicular to AB.) Thus, for any point C in Γ, there is a point X on AB whose distance to C is equal to the average of the distances from X to A and B.

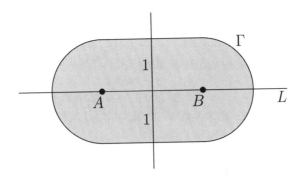

FIGURE 13

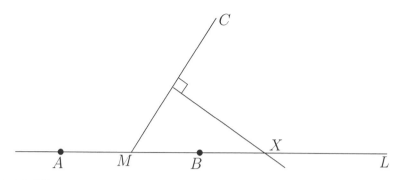

FIGURE 14

Suppose that C is not in Γ and that MC is not perpendicular to AB. Then the perpendicular bisector of MC is not parallel to AB; let X be the intersection point of this perpendicular bisector with L.

If X were in the line segment AB, we would have $CX = MX \leq 1$ and thus C would be in Γ. Therefore, X is outside the line segment AB, and since the distance from X to C equals the distance from X to M, this distance equals the average of the distances from X to A and B.

Thus, the only points C in the plane for which there *might* not be a suitable point X on the line are the points C on the two rays consisting of those points on the perpendicular bisector of AB which are not in Γ. If C is on one of these two open rays, there is no point X: X cannot be on AB because then C would have to be in Γ, and X cannot be outside AB because then X would have to be on the perpendicular bisector of MC, which is parallel to L and thus does not intersect L.

Solution 2. (Bruce Reznick, University of Illinois)   We can assume that L is the x-axis, $A = (-1, 0)$ and $B = (1, 0)$. Let $C = (a, b)$, and by symmetry, suppose $a \geq 0$ and $b \geq 0$. The average of the distances of a point $X = (u, 0)$ on the x-axis to A and B is $\max\{1, |u|\}$; this equals the distance from X to C if and only if $(a - u)^2 + b^2 = \max\{1, |u|^2\}$.

There are two cases. First, consider the equation $(a - u)^2 + b^2 = u^2$. If $a \neq 0$, it has the unique solution $u = (a^2 + b^2)/2a$. This gives a valid X if $|u| \geq 1$; that is, if $a^2 + b^2 \geq 2a$, or $(a - 1)^2 + b^2 \geq 1$. If $a = 0$, the equation has a solution with $|u| \geq 1$ if and only if $b = 0$. Thus, we can find X for C unless C is on the positive y-axis or inside the circle with center $(1, 0)$ and radius 1.

Next, consider the equation $(a - u)^2 + b^2 = 1$. It has a solution if and only if $b \leq 1$, in which case its solutions are $u = a \pm \sqrt{1 - b^2}$. The solution with the smaller absolute value is $a - \sqrt{1 - b^2}$. This leads to an X for C provided $a - \sqrt{1 - b^2} \leq 1$. If $a \leq 1$, this holds for all $b \leq 1$; in particular, it holds if C is on the positive y-axis with $b \leq 1$. If $a > 1$, this holds if and only if $(a - 1)^2 + b^2 \leq 1$; that is, if C is inside the circle with center $(1, 0)$ and radius 1.

We conclude that the only points C for which no X exists are the points $(0, b)$, $|b| > 1$.

Problem 40

Show that there exists a positive number λ such that

$$\int_0^\pi x^\lambda \sin x \, dx = 3.$$

Solution 1. We first observe that

$$\int_0^\pi \sin x \, dx = 2 \quad (\lambda = 0) \qquad \text{and} \qquad \int_0^\pi x \sin x \, dx = \pi \quad (\lambda = 1),$$

the latter integral computed using integration by parts. Since $2 < 3 < \pi$, the Intermediate Value Theorem will imply the existence of a λ between 0 and 1 for which

$$\int_0^\pi x^\lambda \sin x \, dx = 3,$$

provided we prove

$$f(\alpha) = \int_0^\pi x^\alpha \sin x \, dx$$

defines a continuous function on $[0, 1]$.

Suppose $\beta \geq \alpha \geq 0$. Note that $\sin x \geq 0$ on $[0, \pi]$. Further noting that $x^\beta \leq x^\alpha$ for $x \leq 1$, while $x^\beta \geq x^\alpha$ for $x \geq 1$, we have

$$|f(\beta) - f(\alpha)| = \left| \int_0^\pi (x^\beta - x^\alpha) \sin x \, dx \right|$$

$$\leq \int_0^\pi |x^\beta - x^\alpha| \sin x \, dx$$

$$= \int_0^1 (x^\alpha - x^\beta) \sin x \, dx + \int_1^\pi (x^\beta - x^\alpha) \sin x \, dx.$$

Since $|\sin x| \le 1$, we obtain

$$|f(\beta) - f(\alpha)| \le \int_0^1 (x^\alpha - x^\beta)\, dx + \int_1^\pi (x^\beta - x^\alpha)\, dx$$

$$= \frac{\pi^{\beta+1} - 2}{\beta + 1} - \frac{\pi^{\alpha+1} - 2}{\alpha + 1}.$$

This shows that, whether or not $\beta \ge \alpha$, we have

$$|f(\beta) - f(\alpha)| \le \left| \frac{\pi^{\beta+1} - 2}{\beta + 1} - \frac{\pi^{\alpha+1} - 2}{\alpha + 1} \right|.$$

Since the right-hand side of the above tends to 0 as $\beta \to \alpha$, we have

$$\lim_{\beta \to \alpha} f(\beta) = f(\alpha),$$

hence f is continuous on $[0, 1]$, as claimed. The desired λ must therefore exist.

Solution 2. This and the next solution provide two alternative proofs that the function f defined above is continuous on $[0, 1]$.

Note that $\sin x \le x$ for $x \ge 0$ (this can be shown by looking at the derivative of $x - \sin x$). Therefore,

$$|f(\beta) - f(\alpha)| = \left| \int_0^\pi (x^\beta - x^\alpha) \sin x\, dx \right|$$

$$\le \int_0^\pi |x^\beta - x^\alpha| \sin x\, dx$$

$$\le \int_0^\pi |x^{\beta+1} - x^{\alpha+1}|\, dx.$$

For any x with $0 < x \le \pi$, the Mean Value Theorem applied to the function $g(y) = x^{y+1}$ implies that there exists a γ (which depends on x) between α and β for which

$$|x^{\beta+1} - x^{\alpha+1}| = x^{\gamma+1} |\ln x|\, |\beta - \alpha|.$$

It is well known that $\lim_{x \to 0+} x \ln x = 0$ (one can use l'Hôpital's rule). This is enough to imply $x^{\gamma+1} |\ln x|$ is bounded for $0 < x \le \pi$, $0 \le \gamma \le 1$, and hence, for some constant C,

$$|x^{\beta+1} - x^{\alpha+1}| \le C\, |\beta - \alpha|.$$

Therefore,

$$|f(\beta) - f(\alpha)| \le \int_0^\pi C\, |\beta - \alpha|\, dx = C\pi |\beta - \alpha|.$$

This clearly implies $\lim_{\beta \to \alpha} f(\beta) = f(\alpha)$, hence the continuity of f.

Solution 3.   For a third proof of the continuity of f, again begin with

$$|f(\beta) - f(\alpha)| \le \int_0^\pi |x^{\beta+1} - x^{\alpha+1}| \, dx.$$

Because x^{y+1} is continuous on the closed, bounded rectangle $0 \le x \le \pi$, $0 \le y \le 1$, it is uniformly continuous on this rectangle. Given $\varepsilon > 0$, there exists a $\delta > 0$ such that for points (x_1, β) and (x_2, α) in the rectangle with distance between them less than δ, $|x_1^{\beta+1} - x_2^{\alpha+1}| < \varepsilon/\pi$. In particular, for the case $x_1 = x_2 = x$, we see that if $|\beta - \alpha| < \delta$, then

$$|f(\beta) - f(\alpha)| < \int_0^\pi \frac{\varepsilon}{\pi} \, dx = \varepsilon,$$

proving the continuity of f.

Problem 41

What is the fifth digit from the end (the ten thousands digit) of the number $5^{5^{5^{5^5}}}$?

Idea.   Remainder problems generally exhibit some sort of periodic behavior.

Answer.   The fifth digit from the end is 0.

Solution 1.   We are interested in the remainder when $5^{5^{5^{5^5}}}$ is divided by $100{,}000 = 2^5 5^5$. Clearly 5^r is divisible by 5^5 for $r \ge 5$. We determine those r for which $5^r - 5^5$ is divisible by 2^5 as well, implying 5^r and 5^5 have the same last five digits. This will be the case if and only if $5^{r-5} - 1$ is divisible by 32. The binomial expansion of $(1 + 4)^{r-5}$ starts out

$$1 + (r - 5) \, 4 + \frac{(r - 5)(r - 6)}{2} \, 4^2 + \cdots,$$

and all further terms are divisible by 32. Thus, $5^{r-5} - 1$ is divisible by 32 if and only if

$$(r - 5) + \frac{(r - 5)(r - 6)}{2} \, 4$$

is divisible by 8. This certainly occurs for r of the form $8k + 5$. Since, for any m, $(2m + 1)^2 = 4m(m + 1) + 1$, 5 to any even power is of the form $8k + 1$

and 5 to any odd power is of the form $8k + 5$. In particular, $5^{5^{5^{5^5}}}$ is of the form $8k + 5$, so it has the same last five digits as $5^5 = 3125$. Thus, the fifth digit from the end is 0.

Solution 2. We make repeated use of Euler's theorem, which states that if a is an integer relatively prime to $n > 1$, then $a^{\phi(n)} \equiv 1 \pmod{n}$, where $\phi(n)$ is the number of positive integers which are less than or equal to n and relatively prime to n.

As in Solution 1, we are interested in the remainder when $5^{5^{5^{5^5}}}$ is divided by $100,000 = 2^5 5^5$. Because it is divisible by 5^5, we need only determine the remainder upon division by 2^5. Since $\phi(2^5) = 16$, we want the remainder when $5^{5^{5^5}}$ is divided by 16. Since $\phi(16) = 8$, we want the remainder when 5^{5^5} is divided by 8. We now note that every odd square is congruent to 1 (mod 8), hence every odd power of 5 is congruent to 5 modulo 8. In particular,

$$5^{5^5} \equiv 5 \pmod 8.$$

Working backward, we see that

$$5^{5^{5^5}} \equiv 5^5 = 3125 \equiv 5 \pmod{16}.$$

Finally, we conclude

$$5^{5^{5^{5^5}}} \equiv 5^5 \pmod{32}.$$

Rather than computing the remainder upon dividing 5^5 by 32, we observe that

$$5^{5^{5^{5^5}}} \equiv 5^5 \pmod{2^5} \quad \text{and} \quad \pmod{5^5},$$

hence (mod 10^5). Since $5^5 = 3125$, the fifth digit from the end of $5^{5^{5^{5^5}}}$ is a 0.

Problem 42

Describe the set of all points P in the plane such that exactly two tangent lines to the curve $y = x^3$ pass through P.

Solution. We will show that all points P on exactly two tangent lines are of the form $(a, 0)$ or (a, a^3), where $a \neq 0$.

Let $P = (a, b)$. If a tangent to the curve $y = x^3$ passes through P, then the point of tangency (x, x^3) satisfies

$$x^3 - b = 3x^2(x - a),$$

or

$$2x^3 - 3ax^2 + b = 0.$$

If the tangent lines through distinct points (x_1, x_1^3) and (x_2, x_2^3) have the same slope, then we must have $x_1 = -x_2$. However, the tangent line through (x_1, x_1^3) is

$$y = 3x_1^2 x - 2x_1^3,$$

while the one through $(-x_1, -x_1^3)$ is

$$y = 3x_1^2 x + 2x_1^3.$$

Thus, no line is tangent to $y = x^3$ at two points. We conclude that there are exactly two tangents to the curve which pass through P if and only if

$$p(x) = 2x^3 - 3ax^2 + b$$

has exactly two real roots.

Since a cubic polynomial with two real roots must have three, not necessarily distinct, real roots, $p(x)$ must have a multiple root. This occurs if and only if $p(x)$ and $p'(x)$ have a common root. Since $p'(x) = 6x^2 - 6ax$, this common root must be either 0 or a. If $p(0) = 0$, then $b = 0$. The third root of $p(x)$ is then $3a/2$, which is a distinct real root provided $a \neq 0$. On the other hand, if $p(a) = 0$, then $b = a^3$. The third root of $p(x)$ is now $-a/2$, hence again $a \neq 0$. Thus, the only points on exactly two tangent lines to $y = x^3$ are of the form $(a, 0)$ or (a, a^3), where $a \neq 0$.

Problem 43

Show that if $p(x)$ is a polynomial of odd degree greater than 1, then through any point P in the plane, there will be at least one tangent line to the curve $y = p(x)$. Is this still true if $p(x)$ is of even degree?

Solution. If $y = p(x)$, where $p(x)$ is a polynomial of odd degree $d > 1$, then $P = (a, b)$ is on some tangent to the curve if and only if the equation

$$p(x) - b = p'(x)(x - a)$$

has a real solution. Thus, we are looking for a real root of

$$xp'(x) - p(x) - ap'(x) + b,$$

which has degree d (with leading coefficient $(d-1)$ times the leading coefficient of $p(x)$). Since, by the Intermediate Value Theorem, any real polynomial of odd degree has a real root, there is a real x_0 for which the tangent to $y = p(x)$ at $(x_0, p(x_0))$ passes through $P = (a, b)$.

The result does not hold for polynomials of even degree. For instance, since $y = x^2$ is concave up, no tangent line can pass through any point above this parabola.

Comment. For any polynomial $p(x)$ of even degree, one can find points in the plane through which no tangent line to $y = p(x)$ passes. In fact, any point "inside" the curve which is sufficiently far from the x-axis will do. To show this, use the concavity of the curve for $|x|$ large and the boundedness of $p'(x)$ on a finite interval.

Problem 44

Find a solution to the system of simultaneous equations

$$\begin{cases} x^4 - 6x^2y^2 + y^4 = 1 \\ 4x^3y - 4xy^3 = 1, \end{cases}$$

where x and y are real numbers.

Idea. The coefficients and the powers of x and y in the equations are reminiscent of the binomial expansion

$$(x + y)^4 = x^4 + 4x^3y + 6x^2y^2 + 4xy^3 + y^4.$$

Solution. One solution is

$$x = \frac{\sqrt{2 + \sqrt{2 + \sqrt{2}}}}{2^{7/8}}, \quad y = \frac{\sqrt{2 - \sqrt{2 + \sqrt{2}}}}{2^{7/8}}.$$

If we put $z = x + iy$, where $i^2 = -1$, we have

$$z^4 = x^4 + 4x^3(iy) + 6x^2(iy)^2 + 4x(iy)^3 + (iy)^4$$
$$= (x^4 - 6x^2y^2 + y^4) + (4x^3y - 4xy^3)\,i.$$

Hence, the two given equations are equivalent to the one equation $z^4 = 1 + i$ in the complex unknown z. Now write z and $1 + i$ in polar form:

$$z = r(\cos\theta + i\sin\theta), \qquad 1 + i = \sqrt{2}\left(\cos(\pi/4) + i\sin(\pi/4)\right).$$

Using the formula $(\cos\theta + i\sin\theta)^n = \cos n\theta + i\sin n\theta$ (de Moivre's theorem), our equation becomes

$$r^4\left(\cos 4\theta + i\sin 4\theta\right) = \sqrt{2}\left(\cos(\pi/4) + i\sin(\pi/4)\right).$$

We get a solution by taking $r = \sqrt[4]{\sqrt{2}}$, $4\theta = \pi/4$, or equivalently, $r = \sqrt[8]{2}$, $\theta = \pi/16$, and therefore $x = \sqrt[8]{2}\cos(\pi/16)$, $y = \sqrt[8]{2}\sin(\pi/16)$. Using half-angle formulas, this solution can be written as shown above.

Comments. There are exactly three other solutions, obtained by taking

$$4\theta = \pi/4 + 2\pi, \qquad 4\theta = \pi/4 + 4\pi, \qquad 4\theta = \pi/4 + 6\pi.$$

For a more elementary (but laborious) approach, rewrite the equations as

$$(x^2 - y^2)^2 - 4x^2 y^2 = 1,$$

$$4xy(x^2 - y^2) = 1.$$

Then let $xy = u$, $x^2 - y^2 = v$, so the system becomes $v^2 - 4u^2 = 1$, $4uv = 1$. Although this reduces to a quadratic equation in u^2, considerable computation is needed to eventually recover x and y.

Problem 45

Call a convex pentagon "parallel" if each diagonal is parallel to the side with which it does not have a vertex in common. That is, $ABCDE$ is parallel if the diagonal AC is parallel to the side DE and similarly for the other four diagonals. It is easy to see that a regular pentagon is parallel, but is a parallel pentagon necessarily regular?

Answer. No, a parallel pentagon need not be a regular pentagon.

Solution 1. A one-to-one linear transformation of the plane onto itself takes parallel lines to parallel lines. So, start with a regular pentagon, say with vertices

$$\mathbf{v}_k = \begin{pmatrix} \cos(2k\pi/5) \\ \sin(2k\pi/5) \end{pmatrix}, \qquad k = 0, 1, 2, 3, 4,$$

and now simply change the scale on the x- and y-axes. For example, take the specific linear transformation $\mathbf{A} = \left(\begin{smallmatrix} 2 & 0 \\ 0 & 1 \end{smallmatrix}\right)$. This stretches the x-coordinate by a factor of 2 and leaves the y-coordinate unchanged. The resulting figure is a non-regular parallel pentagon.

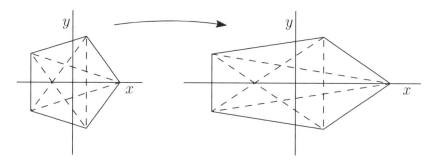

FIGURE 15

Solution 2. Another way to construct a non-regular parallel pentagon is as follows: Start with a square $ABXE$, say of side 1. Extend EX by x units to C and BX by x units to D, where $x > 0$ will be determined later. We will choose x so that $ABCDE$ is a parallel pentagon; obviously it will be non-regular.

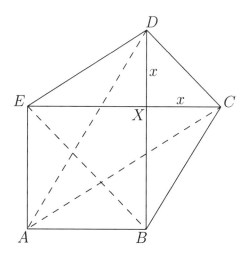

FIGURE 16

Regardless of the value of $x > 0$, $EC \parallel AB$ and $BD \parallel AE$. Also, since $DX = CX = x$, DXC and BXE are both isosceles right triangles, so $BE \parallel CD$. This leaves us to choose x (if possible) so that $AC \parallel DE$ and $AD \parallel BE$. By symmetry about AX, it is enough to find x so that $AC \parallel DE$. For this, it is sufficient to choose x so that triangles DEX and ACE are similar, or equivalently, so that the ratios of corresponding legs of the right triangles are equal to each other. Thus, we want

$$\frac{x}{1} = \frac{DX}{EX} = \frac{AE}{CE} = \frac{1}{1+x}.$$

This holds, for $x > 0$, when $x = (-1 + \sqrt{5})/2$, and thus our construction can be carried out.

Problem 46

Find the sum of the infinite series

$$\sum_{n=1}^{\infty} \frac{1}{2n^2 - n} = 1 + \frac{1}{6} + \frac{1}{15} + \frac{1}{28} + \cdots.$$

Answer. The series sums to $2 \ln 2$.

Solution 1. We begin with the partial fraction decomposition

$$\frac{1}{2n^2 - n} = \frac{2}{2n - 1} - \frac{1}{n} = \frac{2}{2n - 1} - \frac{2}{2n}.$$

Thus,

$$\sum_{n=1}^{\infty} \frac{1}{2n^2 - n} = 2 \sum_{n=1}^{\infty} \left(\frac{1}{2n - 1} - \frac{1}{2n} \right) = 2 \left(1 - \frac{1}{2} + \frac{1}{3} - \frac{1}{4} + \cdots \right).$$

The last step is legitimate because the alternating series on the right is convergent. We now recall the well-known Taylor series expansion

$$\ln(1 + x) = \sum_{n=1}^{\infty} (-1)^{n+1} \frac{x^n}{n},$$

which is valid for $-1 < x \leq 1$. In particular, setting $x = 1$ yields

$$\sum_{n=1}^{\infty} \frac{1}{2n^2 - n} = 2 \ln 2.$$

Solution 2. We express the series as a definite integral we can evaluate. Observing that

$$\frac{1}{2n^2 - n} = \frac{1}{2n - 1} \cdot \frac{1}{n},$$

we have

$$\sum_{n=1}^{\infty} \frac{1}{2n^2 - n} = \sum_{n=1}^{\infty} \frac{1}{n} \int_0^1 x^{2n-2}\, dx.$$

At this point we would like to switch summation and integration. However, we cannot be too cavalier, because $\sum_{n=1}^{\infty} x^{2n-2}/n$ diverges for $x = 1$, making the resulting integral improper. To avoid this difficulty, we can argue as follows.

By the continuity of a power series on its interval of convergence (Abel's limit theorem),

$$\lim_{b \to 1^-} \sum_{n=1}^{\infty} \frac{b^{2n-1}}{2n^2 - n} = \sum_{n=1}^{\infty} \frac{1}{2n^2 - n},$$

and therefore

$$\sum_{n=1}^{\infty} \frac{1}{2n^2 - n} = \lim_{b \to 1^-} \sum_{n=1}^{\infty} \frac{1}{n} \int_0^b x^{2n-2}\, dx = \lim_{b \to 1^-} \int_0^b \sum_{n=1}^{\infty} \frac{x^{2n-2}}{n}\, dx.$$

Now, from the Taylor series for $\ln(1 + x)$, we see that

$$\sum_{n=1}^{\infty} \frac{x^{2n-2}}{n} = \frac{1}{x^2} \sum_{n=1}^{\infty} \frac{x^{2n}}{n} = -\frac{\ln(1 - x^2)}{x^2}, \qquad 0 < x < 1,$$

and so we have

$$\sum_{n=1}^{\infty} \frac{1}{2n^2 - n} = \lim_{b \to 1^-} \int_0^b -\frac{\ln(1 - x^2)}{x^2}\, dx = \int_0^1 -\frac{\ln(1 - x^2)}{x^2}\, dx.$$

An integration by parts followed by a partial fraction decomposition yields

$$\int -\frac{\ln(1 - x^2)}{x^2}\, dx = \frac{\ln(1 - x^2)}{x} - \ln(1 - x) + \ln(1 + x) + C$$

$$= \frac{(1 - x)\ln(1 - x)}{x} + \frac{(1 + x)\ln(1 + x)}{x} + C.$$

The well-known limits

$$\lim_{c \to 0^+} c \ln c = 0 \qquad \text{and} \qquad \lim_{c \to 0} \frac{\ln(1 + c)}{c} = 1$$

(provable using l'Hôpital's rule) imply

$$\sum_{n=1}^{\infty} \frac{1}{2n^2 - n} = \lim_{b \to 1^-} \left[\frac{(1-b)\ln(1-b)}{b} + \frac{(1+b)\ln(1+b)}{b} \right]$$

$$- \lim_{a \to 0^+} \left[\frac{(1-a)\ln(1-a)}{a} + \frac{(1+a)\ln(1+a)}{a} \right]$$

$$= (0 + 2\ln 2) - (-1 + 1) = 2\ln 2.$$

Comment. The second solution is more difficult, but it illustrates a useful method.

Problem 47

For any vector $\mathbf{v} = (x_1, \ldots, x_n)$ in $\mathbf{R}^n$ and any permutation σ of $1, 2, \ldots, n$, define $\sigma(\mathbf{v}) = (x_{\sigma(1)}, \ldots, x_{\sigma(n)})$. Now fix $\mathbf{v}$ and let V be the span of $\{\sigma(\mathbf{v}) \mid \sigma \text{ is a permutation of } 1, 2, \ldots, n\}$. What are the possibilities for the dimension of V?

Solution. The possibilities for $\dim V$ are $0, 1, n-1$, and n.

To see this, first consider the case when all coordinates of $\mathbf{v}$ are equal: $\mathbf{v} = (z, z, \ldots, z)$. Then $\sigma(\mathbf{v}) = \mathbf{v}$ for every permutation σ, so V is just the span of $\mathbf{v}$, which has dimension 0 or 1 according to whether $\mathbf{v}$ is $\mathbf{0}$ or not.

Now suppose not all coordinates of $\mathbf{v}$ are equal; let x and y, with $x \neq y$, be among the coordinates of $\mathbf{v}$. Then we can find permutations σ_1 and σ_2 such that $\sigma_1(\mathbf{v}) = (x, y, a_3, \ldots, a_n)$ and $\sigma_2(\mathbf{v}) = (y, x, a_3, \ldots, a_n)$ for some $a_3, \ldots, a_n \in \mathbf{R}$. Therefore,

$$\frac{1}{y-x}(\sigma_1(\mathbf{v}) - \sigma_2(\mathbf{v})) = (-1, 1, 0, \ldots, 0)$$

is in V. That is, $\mathbf{e}_2 - \mathbf{e}_1 \in V$, where $\mathbf{e}_1, \mathbf{e}_2, \ldots, \mathbf{e}_n$ is the standard basis for $\mathbf{R}^n$. Similarly, $\mathbf{e}_3 - \mathbf{e}_1, \ldots, \mathbf{e}_n - \mathbf{e}_1$ are all in V. It is easy to see that the vectors $\mathbf{e}_2 - \mathbf{e}_1, \mathbf{e}_3 - \mathbf{e}_1, \ldots, \mathbf{e}_n - \mathbf{e}_1$ are linearly independent, so $\dim V \geq n - 1$.

Finally, we can write

$$\mathbf{v} = x_1\mathbf{e}_1 + x_2\mathbf{e}_2 + \cdots + x_n\mathbf{e}_n$$

$$= (x_1 + x_2 + \cdots + x_n)\mathbf{e}_1 + x_2(\mathbf{e}_2 - \mathbf{e}_1) + \cdots + x_n(\mathbf{e}_n - \mathbf{e}_1). \quad (*)$$

This shows that if $x_1 + x_2 + \cdots + x_n = 0$, then $\mathbf{v}$ is in the span of $\mathbf{e}_2 - \mathbf{e}_1, \ldots, \mathbf{e}_n - \mathbf{e}_1$; similarly, each $\sigma(\mathbf{v})$ will be in this span, so V will equal this span and $\dim V = n - 1$. On the other hand, if $x_1 + x_2 + \cdots + x_n \neq 0$, then $(*)$ shows that $\mathbf{e}_1 \in V$ and thus also $\mathbf{e}_2, \ldots, \mathbf{e}_n \in V$, so $V = \mathbf{R}^n$ and $\dim V = n$.

Problem 48

Suppose three circles, each of radius 1, go through the same point in the plane. Let A be the set of points which lie inside at least two of the circles. What is the smallest area A can have?

Solution. The smallest area is $\pi - \frac{3}{2}\sqrt{3}$.

Let C_1, C_2, C_3 be the circles, let O_1, O_2, O_3 be their centers, and let P be their common point. Let A_1, A_2, A_3 be the sets of points inside both C_2 and C_3, inside both C_1 and C_3, and inside both C_1 and C_2, respectively, so that A is the (not necessarily disjoint) union of A_1, A_2 and A_3. Note that the three centers of the circles all have distance 1 to P and therefore lie on a fourth circle with center at P. We may assume that O_2 follows O_1 as one moves counterclockwise around this circle. Let α, β, γ be the counterclockwise angles $O_2PO_3, O_3PO_1,$ O_1PO_2, respectively. Then $\alpha + \beta + \gamma = 2\pi$, and we may assume $\alpha \geq \beta, \alpha \geq \gamma$. There are then two cases, as illustrated by Figures 17 and 18.

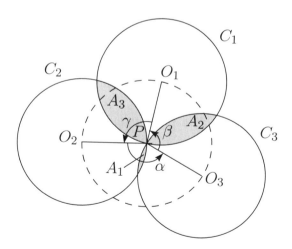

FIGURE 17
Case 1. $\alpha \leq \pi$.

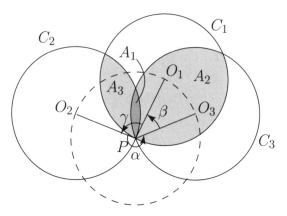

FIGURE 18
Case 2. $\alpha > \pi$.

In Case 1, there is no overlap between the regions A_1, A_2, A_3, so the area of A is the sum of the areas of A_1, A_2, and A_3. In Case 2, on the other hand, A_1 is the intersection of A_2 and A_3, so the area of A equals the sum of the areas of A_2 and A_3 minus the area of A_1.

To find the areas of the individual A_i, it is enough to find how the area of A_2 depends on β. By symmetry, the result of this computation will immediately give us all the other areas, since the region A_1 in Case 2 (for the angle $\alpha > \pi$) is congruent to the region A_1 in Case 1 for the angle $2\pi - \alpha$.

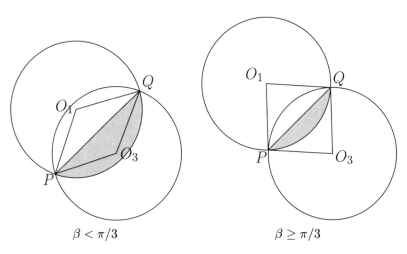

$\beta < \pi/3$ $\qquad\qquad\qquad$ $\beta \geq \pi/3$

FIGURE 19

The area of A_2 is twice the area of the shaded region shown in Figure 19. That region is obtained by removing triangle O_1PQ from circular sector O_1PQ, where Q is the second intersection point of C_1 and C_3. Now the area of the circular sector O_1PQ is $(\pi - \beta)/2$ and the area of triangle O_1PQ is

$$\frac{1}{2}\sin(\pi - \beta) = \frac{1}{2}\sin\beta.$$

Thus, the area of the shaded region is $(\pi - \beta - \sin\beta)/2$ and the area of A_2 is $\pi - \beta - \sin\beta$.

Combining the areas as indicated above, we find that in both cases, the area we wish to minimize is

$$\pi - \sin\alpha - \sin\beta - \sin\gamma, \qquad \alpha, \beta, \gamma \geq 0, \qquad \alpha + \beta + \gamma = 2\pi.$$

Suppose the angles are not all equal. Then $\alpha > 2\pi/3$ and at least one of β and γ is less than $2\pi/3$. Suppose $\beta < 2\pi/3$. Now fix γ and note that

$$\sin\alpha + \sin\beta = 2\sin\left(\frac{\alpha + \beta}{2}\right)\cos\left(\frac{\alpha - \beta}{2}\right)$$

$$= 2\sin\left(\pi - \frac{\gamma}{2}\right)\cos\left(\frac{\alpha - \beta}{2}\right).$$

Since $0 \leq \gamma \leq 2\pi$, $\sin(\pi - \frac{\gamma}{2}) \geq 0$. Now steadily decrease α and increase β, in such a way that their sum remains constant, until one or the other is equal to $2\pi/3$. As the difference $\alpha - \beta$ decreases, the value of $\sin\alpha + \sin\beta$ increases, so the area decreases. Thus we can assume that one of the angles is equal to $2\pi/3$. Suppose $\beta = 2\pi/3$ (the other case is similar). Then

$$\sin\alpha + \sin\gamma = 2\sin\left(\pi - \frac{\beta}{2}\right)\cos\left(\frac{\alpha - \gamma}{2}\right) = \sqrt{3}\cos\left(\frac{\alpha - \gamma}{2}\right).$$

This sum is maximized when $\cos(\frac{\alpha - \gamma}{2}) = 1$, or equivalently, when

$$\alpha = \gamma = 2\pi/3,$$

as well. Thus, under the given constraints, $\sin\alpha + \sin\beta + \sin\gamma$ is maximized, and the area of A is minimized, when α, β, γ all equal $2\pi/3$, and the minimal area is $\pi - 3\sin(2\pi/3) = \pi - \frac{3}{2}\sqrt{3}$.

Comment. As we have seen, the problem reduces to that of finding the maximum value of

$$\sin\alpha + \sin\beta + \sin\gamma, \qquad \alpha, \beta, \gamma > 0, \qquad \alpha + \beta + \gamma = 2\pi.$$

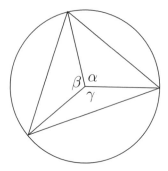

Figure 20 shows how this can be reformulated geometrically as: Given a circle of radius 1, find the maximum area of an inscribed triangle. It is known that the area of such a triangle is maximized when the triangle is equilateral, that is, when $\alpha = \beta = \gamma = 2\pi/3$.

Problem 49

How many real solutions does the equation $\sqrt[7]{x} - \sqrt[5]{x} = \sqrt[3]{x} - \sqrt{x}$ have?

Answer. There are three solutions to the given equation.

Solution 1. We wish to find the number of solutions of

$$\sqrt{x} - \sqrt[3]{x} - \sqrt[5]{x} + \sqrt[7]{x} = 0.$$

Letting $x = y^{210}$, it is enough to determine the number of nonnegative roots of

$$y^{105} - y^{70} - y^{42} + y^{30},$$

which factors as

$$y^{30}(y - 1)\left[y^{40}\left(y^{34} + y^{33} + \cdots + y + 1\right) - \left(y^{11} + y^{10} + \cdots + y + 1\right)\right].$$

If we let

$$p(y) = y^{40}\left(y^{34} + y^{33} + \cdots + y + 1\right) - \left(y^{11} + y^{10} + \cdots + y + 1\right),$$

then $p(0) = -1$, while $p(1) = 23$. Thus, by the Intermediate Value Theorem, there must be a root of $p(y)$ between 0 and 1. On the other hand, the coefficients of $p(y)$ change sign exactly once, hence, by Descartes' rule of signs, $p(y)$ can have at most one positive root. Therefore, there are three values of x, namely 0, 1, and some number between 0 and 1, which solve the given equation.

Solution 2. Let

$$f(x) = \sqrt[7]{x} - \sqrt[5]{x} - \sqrt[3]{x} + \sqrt{x},$$

which has domain $x \geq 0$. By inspection, $f(0) = f(1) = 0$. Differentiating f, we obtain

$$f'(x) = \frac{1}{7}x^{-6/7} - \frac{1}{5}x^{-4/5} - \frac{1}{3}x^{-2/3} + \frac{1}{2}x^{-1/2}.$$

In particular, $f'(1) = 23/210 > 0$, hence $f(x) < 0$ for x less than 1 and sufficiently close to 1. Since

$$\lim_{x \to 0+} x^{6/7} f'(x) = \frac{1}{7},$$

$f'(x) > 0$ for sufficiently small positive x. For such x, the Mean Value Theorem implies $f(x) = f'(c)x$ for some c, $0 < c < x$, hence $f(x) > 0$. The continuity of f now implies the existence of a zero of f in the open interval $(0, 1)$.

If we can show that $f'(x)$, or equivalently, $x^{6/7} f'(x)$, has at most 2 positive zeros, then Rolle's theorem will imply that $f(x)$ has at most 3 nonnegative zeros. We prove the more general result that

$$g(x) = \sum_{j=1}^{n} a_j x^{r_j}, \qquad r_1 > r_2 > \cdots > r_n, \quad a_j \neq 0,$$

can have at most $n-1$ positive roots. This is clear for $n = 1$. Assume the result for a sum of $n - 1$ terms. Differentiating $x^{-r_n} g(x)$ yields

$$\sum_{j=1}^{n-1} (r_j - r_n) a_j x^{r_j - r_n - 1},$$

which has at most $n - 2$ positive roots by the inductive hypothesis. Rolle's theorem implies $x^{-r_n} g(x)$, and therefore also $g(x)$, has at most $n - 1$ positive roots, proving the generalization.

It follows that $f(x)$ has exactly three zeros.

Problem 50

Let $\mathbf{A} \neq \mathbf{0}$ and $\mathbf{B}_1, \mathbf{B}_2, \mathbf{B}_3, \mathbf{B}_4$ be 2×2 matrices (with real entries) such that

$$\det(\mathbf{A} + \mathbf{B}_i) = \det \mathbf{A} + \det \mathbf{B}_i \qquad \text{for } i = 1, 2, 3, 4.$$

Show that there exist real numbers k_1, k_2, k_3, k_4, not all zero, such that

$$k_1 \mathbf{B}_1 + k_2 \mathbf{B}_2 + k_3 \mathbf{B}_3 + k_4 \mathbf{B}_4 = \mathbf{0}.$$

Solution.   Let

$$\mathbf{A} = \begin{pmatrix} a & b \\ c & d \end{pmatrix} \qquad \text{and} \qquad \mathbf{B}_i = \begin{pmatrix} w_i & x_i \\ y_i & z_i \end{pmatrix}, \qquad i = 1, 2, 3, 4.$$

If $\det(\mathbf{A} + \mathbf{B}_i) = \det \mathbf{A} + \det \mathbf{B}_i$, then

$$dw_i - cx_i - by_i + az_i = 0.$$

Because $\mathbf{A} \neq \mathbf{0}$, the solution space to $dw - cx - by + az = 0$ is a 3-dimensional vector space. Since any four vectors in a 3-dimensional space are linearly dependent, there must exist k_1, k_2, k_3, k_4, not all 0, for which

$$k_1 \mathbf{B}_1 + k_2 \mathbf{B}_2 + k_3 \mathbf{B}_3 + k_4 \mathbf{B}_4 = \mathbf{0}.$$

Comment.   The "obvious" analog of this statement for nine 3×3 matrices $\mathbf{B}_1, \ldots, \mathbf{B}_9$ is false.

Problem 51

Two ice fishermen have set up their ice houses on a perfectly circular lake, in exactly opposite directions from the center, two-thirds of the way from the center to the lakeshore. The point of this symmetrical arrangement is that any fish that can be lured will swim toward the closest lure, and therefore both fishermen have equal expectations of their catch. Can a third ice house be put on the lake in such a way that all three fishermen will have equal expectations at least to the extent that the three regions, each consisting of all points on the lake for which one of the three ice houses is closest, all have the same area?

Idea.   If a third ice house were placed near the edge of the lake, it could not "control" one-third of the lake; if it were placed in the center of the lake, it would control more than one-third of the lake. There should be an intermediate spot somewhere between these extremes where the third ice house will control exactly one third of the area.

Solution. Yes, it is possible to put a third ice house on the lake in such a way that the three areas described in the problem are all equal. To see why, suppose that the center of the lake is at C and that the diameter on which the first two ice houses are located is AB; let H_1, H_2 be the locations of these ice houses, so we are given that $H_1C = \frac{2}{3}AC$ and $CH_2 = \frac{2}{3}CB$.

If the third ice house is placed at the center of the lake, the three areas controlled by the three ice houses will be parallel strips of the circle bounded by the perpendicular bisectors of the line segments H_1C and CH_2, and it is easy to see that the third ice house controls more than one-third of the area of the circle.

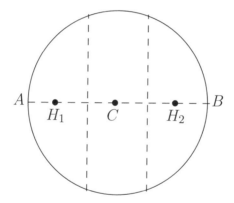

FIGURE 21

Suppose we place the third ice house at the same distance from C as H_1 and H_2, but along a radius at right angles to AB. The shaded sector in Figure 22 is the region of points for which the closest ice house is H_3, and it is easy to see that the sector is one quarter of the circle.

Now imagine moving the third ice house along the radius toward the center of the lake. The three areas (controlled by each of the ice houses) will change continuously as we do this. By symmetry, the regions consisting of points for which H_1 is closest and of points for which H_2 is closest will always have the same area. If we look at the difference between this area and the area of points for which H_3 is closest, we see that this difference is originally positive

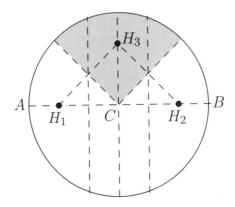

FIGURE 22

(specifically, $3/8 - 1/4 = 1/8$ of the area of the circle) but eventually becomes negative (by the time H_3 reaches the center of the circle). Therefore, by the Intermediate Value Theorem, the difference must be zero at some intermediate point. To get all three areas equal, put the third ice house at that point.

Problem 52

Note that the integers 2, -3, and 5 have the property that the difference of any two of them is an integer times the third. Suppose three distinct integers a, b, c have this property.

a. Show that a, b, c cannot all be positive.
b. Now suppose that a, b, c, in addition to having the above property, have no common factors (except $1, -1$). Is it true that one of the three integers has to be either $1, 2, -1$, or -2?

Solution.   Without loss of generality, we may assume $a < b < c$.

a.   We prove the stronger result that one of the three, in our case a, must be negative. Suppose $0 \le a$. Then $0 < b - a \le b < c$, hence $0 < (b-a)/c < 1$, a contradiction.

b.   One of a, b, c must be ± 1 or ± 2.

Replacing (a, b, c) by $(-c, -b, -a)$ if necessary, we may assume $0 \le b$. Since $c - a$ is a nonzero multiple of b, we must actually have $a < 0 < b < c$. The divisibility condition implies the existence of positive integers m and n such that

$$b - c = na,$$

$$b - a = mc.$$

Then $c - a = mc - na$ (by subtraction), so $(m-1)c = (n-1)a$. This, together with $c > 0$ and $a < 0$, implies $m = 1, n = 1$. Thus, $b - c = a$, or $c + a = b$. Using this, and the supposition that $c - a = kb$ for some integer k, we have

$$2c = (c + a) + (c - a) = b + kb = (k+1)b.$$

If b and c are relatively prime, then b divides 2, and we are done. If not, then the equation $a = b - c$ implies a, b, and c have a nontrivial common factor, a contradiction.

Problem 53

Let $\mathbf{A}$ be an $m \times n$ matrix with every entry either 0 or 1. How many such matrices $\mathbf{A}$ are there for which the number of 1's in each row and each column is even?

Solution.   There are $2^{(m-1)(n-1)}$ such matrices.

To see this, we consider only "0-1" matrices (matrices whose entries are either 0 or 1), and call a row or column of such a matrix *odd* or *even* according to whether it contains an odd or even number of 1's. We then want to find the number of $m \times n$ matrices for which all rows and columns are even.

Let $\mathbf{A}$ be such a matrix. The $(m-1) \times (n-1)$ submatrix in the upper left corner of $\mathbf{A}$ is also a 0-1 matrix. We shall show that any $(m-1) \times (n-1)$ 0-1 matrix $\mathbf{B}$ uniquely determines an $m \times n$ 0-1 matrix $\mathbf{A}$ whose rows and columns are all even.

So let $\mathbf{B}$ be an $(m-1) \times (n-1)$ matrix; we will construct a matrix of the form

$$\mathbf{A} = \begin{pmatrix} & & & a_{1n} \\ & \mathbf{B} & & \vdots \\ & & & a_{m-1,n} \\ a_{m1} & \cdots & a_{m,n-1} & a_{mn} \end{pmatrix}.$$

To make the first row of $\mathbf{A}$ even, a_{1n} must be 0 or 1 depending on whether the first row of $\mathbf{B}$ is even or odd. A similar argument for the other rows and columns of $\mathbf{B}$ shows that $a_{1n}, \ldots a_{m-1,n}$ and $a_{m1}, \ldots, a_{m,n-1}$ are all determined by the entries of $\mathbf{B}$. Now $\mathbf{A}$ will have the desired property provided that the last row and column of **A** come out to be even. This can be arranged by making the right choice (0 or 1) for a_{mn}, provided the parity (even or odd) of the "partial row" $(a_{m1}, \ldots, a_{m,n-1})$ is the same as the parity of the "partial column"

$$\begin{pmatrix} a_{1n} \\ \vdots \\ a_{m-1,n} \end{pmatrix}.$$

Given how we got the entries in this partial row and in this partial column, this means that the parity of the number of odd rows of **B** must be the same as the parity of the number of odd columns of **B**. However, this is always true, because each of these parities is odd or even according to whether the whole matrix **B** contains an odd or even number of 1's. Thus there is exactly one way to complete an $(m-1) \times (n-1)$ matrix **B** to a matrix **A** with the desired property.

Clearly, different $(m-1) \times (n-1)$ matrices **B** lead to different matrices **A** of the desired type. Thus, we see that there is a 1-1 correspondence between $m \times n$ matrices **A** for which all rows and columns are even and arbitrary $(m-1) \times (n-1)$ matrices **B**. Since we can choose each of the $(m-1)(n-1)$ entries of **B** independently, there are $2^{(m-1)(n-1)}$ choices for **B**, and we are done.

Comments. In terms of modular arithmetic, the construction of **A** from **B** can be phrased as follows. Given $\mathbf{B} = (b_{ij})$, set

$$a_{nj} \equiv \sum_{i=1}^{n-1} b_{ij} \pmod{2} \quad \text{and} \quad a_{im} \equiv \sum_{j=1}^{m-1} b_{ij} \pmod{2}.$$

As for a_{nm}, the row condition requires

$$a_{nm} \equiv \sum_{j=1}^{m-1} a_{nj} \equiv \sum_{j=1}^{m-1} \sum_{i=1}^{n-1} b_{ij} \pmod{2},$$

whereas the column condition requires

$$a_{nm} \equiv \sum_{i=1}^{n-1} a_{im} \equiv \sum_{i=1}^{n-1} \sum_{j=1}^{m-1} b_{ij} \pmod{2}.$$

Since these sums are identical, a_{nm} is well defined.

More generally, a similar argument shows that there are $N^{(m-1)(n-1)}$ $m \times n$ matrices, with entries from $\{0, 1, 2, \ldots, N-1\}$, whose rows and columns each add to a multiple of N.

Problem 54

Suppose A and B are convex subsets of $\mathbf{R}^3$. Let C be the set of all points R for which there are points P in A and Q in B such that R lies between P and Q. Does C have to be convex?

Solution. The set C is convex.

We identify points in $\mathbf{R}^3$ with vectors in order to be able to add points and to multiply points by scalars. Let R_1 and R_2 be points of C. We show that every point between R_1 and R_2 is in C by showing that for every t, $0 < t < 1$,

$$R = t\,R_1 + (1-t)\,R_2$$

is in C.

Let R_i, $i = 1, 2$, be between P_i in A and Q_i in B. Then there must be u_i, $0 \le u_i \le 1$, for which

$$R_i = u_i\,P_i + (1 - u_i)\,Q_i.$$

If $u_1 = u_2 = 0$, then $R = t\,Q_1 + (1-t)Q_2$ is in B, hence in C. Similarly, if $u_1 = u_2 = 1$, then R is in A, hence in C. We now assume neither holds. Combining the previous equations, we have

$$R = \big(t\,u_1\,P_1 + (1-t)u_2\,P_2\big) + \big(t(1-u_1)\,Q_1 + (1-t)(1-u_2)\,Q_2\big).$$

To use the convexity of A, the coefficients of P_1 and P_2 must be nonnegative (which they are) and sum to 1; therefore, we will factor $d = t\,u_1 + (1-t)u_2$ from the first bracketed term above. Note that $1 - d = t(1-u_1) + (1-t)(1-u_2)$, so we factor this from the second term, and we have

$$R = d\left(\frac{t u_1}{d}P_1 + \frac{(1-t)u_2}{d}P_2\right) + (1-d)\left(\frac{t(1-u_1)}{1-d}Q_1 + \frac{(1-t)(1-u_2)}{1-d}Q_2\right).$$

(We will see below that $0 < d < 1$, so $d \ne 0$ and $(1-d) \ne 0$.) Now since

$$\frac{t u_1}{d} + \frac{(1-t)u_2}{d} = 1,$$

the convexity of A allows us to conclude that

$$\frac{t u_1}{d}P_1 + \frac{(1-t)u_2}{d}P_2 = P$$

for some P in A. Similarly,

$$\frac{t(1-u_1)}{1-d}Q_1 + \frac{(1-t)(1-u_2)}{1-d}Q_2 = Q$$

for some Q in B. Since $0 < d = tu_1 + (1-t)u_2 < t + (1-t) = 1$,

$$R = dP + (1-d)Q$$

is in C, and we are done.

Problem 55

Suppose we have a configuration (set) of finitely many points in the plane which are not all on the same line. We call a point in the plane a *center* for the configuration if for every line through that point, there is an equal number of points of the configuration on either side of the line.

a. Give a necessary and sufficient condition for a configuration of four points to have a center.
b. Is it possible for a finite configuration of points (not all on the same line) to have more than one center?

Solution. a.   For a configuration of four noncollinear points to have a center, it is necessary and sufficient for the four points to be the vertices of a convex quadrilateral (by which we mean a quadrilateral whose interior angles all measure less than $180°$).

Given a convex quadrilateral, it is clear that the intersection of the two diagonals is a center for the configuration of the four vertices. Conversely, let A, B, C, and D be four noncollinear points with center O. The line AO must contain a second point, say C. The points B and D must be on opposite sides of the line AO. We claim $ABCD$ is a convex quadrilateral. The interior angle at B is an angle of the triangle ABC, hence measures less than $180°$. Similarly, the interior angle at D measures less than $180°$. Since no line contains A, B, C, and D, the line BO must also contain D. Arguing as above, the interior angles of $ABCD$ at A and C must measure less than $180°$. We have proved the quadrilateral is convex.

   b.   No configuration of noncollinear points can have more than one center. Suppose a configuration has center O. Let O' be a second point in the plane. Choose a point A of the configuration not on the line OO'. Since O is a center for the configuration, the line AO must have an equal number of points (of the configuration) on either side of it. Now consider the line L through O' parallel to AO. The side of L containing A has at least one more point (in fact, at least two more points) than the side of AO not containing O'. The other side of L has at most the same number of points as the side of AO containing

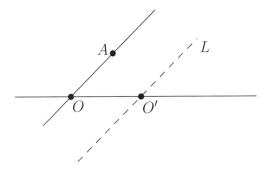

FIGURE 23

O'. Thus, the side of L containing A has more points of the configuration than the other side of L, and O' cannot be a center for the configuration.

Comment. For O to be a center for a configuration of $2n$ points, no three on a line, it is still necessary and sufficient that every line through O and one point contain a second point, with $n - 1$ points on either side of the line. Even with six points, the configuration may not form a convex hexagon, as shown below.

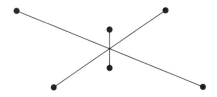

FIGURE 24

Problem 56

Find all real solutions x of the equation

$$x^{10} - x^8 + 8x^6 - 24x^4 + 32x^2 - 48 = 0.$$

Answer. The only real solutions are $\sqrt{2}$ and $-\sqrt{2}$.

Solution 1. Since the powers of x in $x^{10} - x^8 + 8x^6 - 24x^4 + 32x^2 - 48$ are all even, it is sufficient to find all nonnegative solutions of

$$y^5 - y^4 + 8y^3 - 24y^2 + 32y - 48 = 0.$$

We make use of the rational root theorem to search for roots. This theorem states that the only possible rational roots of a polynomial with integer coefficients are quotients of a factor of the constant term (possibly negative) by a factor of the leading coefficient. In our case, it implies any rational root must be an integral factor of 48. The moderate size of the coefficients (as well as ease of computation) leads us to begin with the small factors. We soon see that $y = 2$ is a solution. Long division yields

$$y^5 - y^4 + 8y^3 - 24y^2 + 32y - 48 = (y - 2)(y^4 + y^3 + 10y^2 - 4y + 24).$$

The small size of the lone negative coefficient of the quartic factor above leads us to suspect there are no further nonnegative roots of the polynomial. One of many ways to show this is to rewrite the quartic as

$$y^4 + y^3 + 9y^2 + (y - 2)^2 + 20.$$

This latter expression is clearly at least 20 for $y \geq 0$. We conclude that the only real solutions to the original tenth-degree equation are $\sqrt{2}$ and $-\sqrt{2}$.

Solution 2. As in the first solution, we are led to consider the polynomial

$$y^5 - y^4 + 8y^3 - 24y^2 + 32y - 48.$$

It may be rewritten as

$$y^5 - y^4 + 4 \cdot 2y^3 - 6 \cdot 2^2 y^2 + 4 \cdot 2^3 y - 2^4 - 32, \quad \text{or} \quad y^5 - (y - 2)^4 - 32.$$

We find the root $y = 2$ by inspection. The derivative of the above polynomial is $5y^4 - 4(y - 2)^3$. It is then easy to see that $y^5 - y^4 + 8y^3 - 24y^2 + 32y - 48$ is strictly increasing for all real y (consider the cases $y \leq 2$ and $y \geq 2$ separately), hence there are no other real zeros. Again, this shows the only real solutions to the original tenth-degree equation are $\sqrt{2}$ and $-\sqrt{2}$.

Problem 57

An ordinary die is cubical, with each face showing one of the numbers 1, 2, 3, 4, 5, 6. Each face borders on four other faces; each number is "surrounded" by four of the other numbers. Is it possible to make a die in the shape of a regular dodecahedron, where each of the numbers 1, 2, 3, 4, 5, 6 occurs on two different faces and is "surrounded" both times by all five other numbers? If so, in how many essentially different ways can it be done?

Solution.　Yes, it is possible to have each of the numbers 1, 2, 3, 4, 5, 6 occurring on two of the faces of the dodecahedron, surrounded each time by the other five numbers. Figure 25 shows one way to do this; there are twelve essentially different ways in which it can be done.

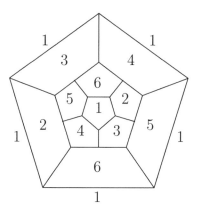

FIGURE **25**

　For the conditions to be satisfied, opposite faces must always have the same number. Thus, a solution is completely determined by the numbers placed on one face and its five neighbors. To count the number of essentially different solutions, we can assume (by rotating the dodecahedron) that the "central" face has a 1; by rotating the dodecahedron around this "central" face, we can further assume that the face represented by the pentagon "northwest" of the central pentagon has a 2 (as shown in Figure 26). At first sight, every placement of the numbers 3, 4, 5, 6 in the other four faces adjacent to the central one will give rise to an essentially different solution. There are $4! = 24$ such placements. But wait! Turning the dodecahedron around so that the front face is turned to the back will replace the front face, labeled 1, and the clockwise pattern surrounding it, labeled $(2, a, b, c, d)$, by the back face, labeled 1, and the mirror image surrounding it, namely $(2, d, c, b, a)$. Thus, each labeled dodecahedron

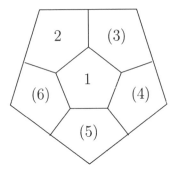

yields two placements of 3, 4, 5, 6 as described above, and therefore the number of essentially different labelings is $24/2 = 12$.

Problem 58

Let k be a positive integer. Find the largest power of 3 which divides $10^k - 1$.

Solution. If $k = 3^m n$, where n is not divisible by 3, we prove that $10^k - 1$ is divisible by 3^{m+2}, but not by 3^{m+3}.

The proof is by induction on m, and uses the fact that an integer is divisible by 3 or 9 if and only if the sum of its digits is divisible by 3 or 9, respectively. For $m = 0$, k is not divisible by 3. We have

$$10^k - 1 = 9 \cdot \underbrace{111 \cdots 1}_{k \text{ digits}},$$

and the sum of the digits of $111 \cdots 1$ is k, so $10^k - 1$ is divisible by 3^2, but not by 3^3. We now assume the claim holds for m. Let $k = 3^{m+1} n$ where n is not divisible by 3. Then

$$10^k - 1 = (10^{3^m n})^3 - 1 = (10^{3^m n} - 1)(10^{2 \cdot 3^m n} + 10^{3^m n} + 1).$$

By the inductive hypothesis, the largest power of 3 dividing the first factor is 3^{m+2}. The sum of the digits of the second factor is 3, which is divisible by 3, but not by 3^2. Therefore, the highest power of 3 dividing $10^k - 1$ is 3^{m+3}, and the proof is complete.

Problem 59

Consider an arbitrary circle of radius 2 in the coordinate plane. Let n be the number of lattice points inside, but not on, the circle.
  a. What is the smallest possible value for n?
  b. What is the largest possible value for n?

Answer. The smallest possible value of n is 9; the largest is 14.

Solution. We begin by making several reductions, in order to simplify the calculations. First, by shifting the origin if necessary, we may assume that no lattice point is closer to the center of the circle than $(0,0)$. By rotating the plane through a multiple of $90°$, we may assume the center lies in the first quadrant or on its boundary. Finally, by reflecting the plane in the line $y = x$ if necessary, we may assume the center of the circle lies in the region R (shaded in Figure 27) defined by $0 \le x \le 1/2, 0 \le y \le x$.

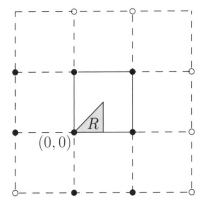

FIGURE 27

Any lattice point outside the square $-1 \le x \le 2, -1 \le y \le 2$ has distance at least two from the unit square $(0 \le x \le 1, 0 \le y \le 1)$, hence cannot be inside the circle. Simple calculations show that the eight lattice points indicated in the figure by "•" have distance less than 2 from *every* point in R. Further computation shows that only the six lattice points indicated in the figure by "∘" have distance less than 2 to some, but not all, of R. We see that $(-1, -1)$ is the only such point whose distance from the origin is less than 2. On the other hand, each point of R *except* the origin has distance less than 2 to $(2, 0)$. Therefore, the smallest possible n is 9, obtained, for instance, when the circle has center at $(0, 0)$.

The largest n is clearly no more than 14, the number of lattice points under consideration. If there is a point in R whose distance from all 14 of the lattice points is less than 2, the symmetry of the problem implies that there is such a point on the line $x = 1/2$. Any point $(1/2, y), 0 \leq y \leq 1/2$, has distance less than 2 to $(2, 0)$ and to $(2, 1)$. The distance to the other four questionable lattice points will be less than 2 if and only if

$$\left(\frac{3}{2}\right)^2 + (y + 1)^2 < 4 \qquad \text{and} \qquad \left(\frac{1}{2}\right)^2 + (2 - y)^2 < 4.$$

These inequalities simplify to

$$y < \frac{\sqrt{7}}{2} - 1 \qquad \text{and} \qquad 2 - \frac{\sqrt{15}}{2} < y.$$

Since

$$0 < 2 - \frac{\sqrt{15}}{2} < \frac{\sqrt{7}}{2} - 1 < \frac{1}{2},$$

there is a y for which the circle of radius 2 and center $(1/2, y)$ encloses 14 lattice points. As one might suspect, one such circle is centered at $(1/2, 1/4)$.

Problem 60

Let a and b be nonzero real numbers and (x_n) and (y_n) be sequences of real numbers. Given that

$$\lim_{n \to \infty} \frac{a x_n + b y_n}{\sqrt{x_n^2 + y_n^2}} = 0$$

and that x_n is never 0, show that

$$\lim_{n \to \infty} \frac{y_n}{x_n}$$

exists and find its value.

Idea. Let $z_n = y_n/x_n$. Note that

$$\frac{a x_n + b y_n}{\sqrt{x_n^2 + y_n^2}} = \pm \frac{a + b z_n}{\sqrt{1 + z_n^2}}.$$

If we assume for the moment that $L = \lim_{n\to\infty} z_n$ exists, then we have

$$0 = \lim_{n\to\infty} \frac{a + bz_n}{\sqrt{1 + z_n^2}} = \frac{a + bL}{\sqrt{1 + L^2}},$$

and so $L = -a/b$.

Solution 1. To show that L exists, we first show that the sequence (z_n) is bounded. We know that

$$\lim_{n\to\infty} \frac{a + bz_n}{\sqrt{1 + z_n^2}} = 0;$$

on the other hand,

$$\lim_{z\to\infty} \frac{a + bz}{\sqrt{1 + z^2}} = b \quad \text{and} \quad \lim_{z\to-\infty} \frac{a + bz}{\sqrt{1 + z^2}} = -b.$$

Now, if (z_n) were unbounded, it would have some subsequence with limit either ∞ or $-\infty$, contradicting the assumption that $b \neq 0$.

Now that we know (z_n) is bounded, say $|z_n| \leq M$ for all n, we have

$$\left| \frac{a + bz_n}{\sqrt{1 + z_n^2}} \right| \geq \left| \frac{a + bz_n}{\sqrt{1 + M^2}} \right|$$

and hence by the squeeze principle,

$$\lim_{n\to\infty} \frac{a + bz_n}{\sqrt{1 + M^2}} = 0,$$

which implies $\lim_{n\to\infty}(a + bz_n) = 0$, and finally $\lim_{n\to\infty} z_n = -a/b$.

Solution 2. (Provided by Meyer Jerison, Purdue University)   With no loss of generality, assume $x_n^2 + y_n^2 = 1$. This amounts to a change of notation:

$$x_n \text{ to } \frac{x_n}{\sqrt{x_n^2 + y_n^2}}, \qquad y_n \text{ to } \frac{y_n}{\sqrt{x_n^2 + y_n^2}}.$$

Now, $ax_n + by_n$ is the dot product of the unit vector (x_n, y_n) with (a, b). The hypothesis is that the dot product tends to 0, so that y_n/x_n should tend to the slope orthogonal to (a, b), namely, $-a/b$. Note that the sequence of unit vectors (x_n, y_n) need not converge; they may flip back and forth.

For an algebraic version of this argument, express the points (x_n, y_n) and (b, a) in polar coordinates: $(1, \theta_n)$ and (A, α). Then

$$ax_n + by_n = A \sin \alpha \cos \theta_n + A \cos \alpha \sin \theta_n = A \sin (\theta_n + \alpha).$$

The hypothesis becomes $\lim_{n\to\infty} \sin(\theta_n + \alpha) = 0$; hence,

$$\lim_{n\to\infty} \tan(\theta_n + \alpha) = 0.$$

This implies

$$\lim_{n\to\infty} \tan \theta_n = \lim_{n\to\infty} \frac{\tan(\theta_n + \alpha) - \tan \alpha}{1 + \tan(\theta_n + \alpha)\tan \alpha} = -\tan \alpha.$$

But $\tan \theta_n = y_n/x_n$ and $\tan \alpha = a/b$.

Problem 61

What is the angle between adjacent faces of the regular icosahedron?

Answer. The angle is $\cos^{-1}(-\sqrt{5}/3) \approx 138.2°$.

Solution. More generally, suppose that AU, AW, and AV are three concurrent lines in $\mathbf{R}^3$, and consider the problem of finding the angle θ between the planes containing triangles UAW and WAV. Let $\alpha = \angle UAW$, $\beta = \angle WAV$, and $\gamma = \angle UAV$ (see Figure 28).

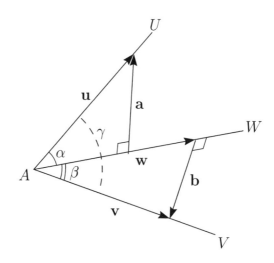

FIGURE 28

Let $\mathbf{u}$, $\mathbf{v}$ be unit vectors in the directions of AU and AV respectively; let $\mathbf{w}$ be the unit vector in the direction of AW. Then, since $\mathbf{u}$, $\mathbf{v}$, and $\mathbf{w}$ are unit vectors, we have

$$\mathbf{u} \cdot \mathbf{w} = \cos \alpha, \qquad \mathbf{v} \cdot \mathbf{w} = \cos \beta, \qquad \mathbf{u} \cdot \mathbf{v} = \cos \gamma.$$

The angle θ is the angle between

$$\mathbf{a} = \mathbf{u} - (\mathbf{u} \cdot \mathbf{w})\mathbf{w} \qquad \text{and} \qquad \mathbf{b} = \mathbf{v} - (\mathbf{v} \cdot \mathbf{w})\mathbf{w}.$$

We have $|\mathbf{a}| = \sin \alpha$, $|\mathbf{b}| = \sin \beta$, and

$$\cos \theta = \frac{\mathbf{a} \cdot \mathbf{b}}{|\mathbf{a}| \, |\mathbf{b}|} = \frac{\mathbf{u} \cdot \mathbf{v} - (\mathbf{u} \cdot \mathbf{w})(\mathbf{v} \cdot \mathbf{w})}{\sin \alpha \sin \beta} = \frac{\cos \gamma - \cos \alpha \cos \beta}{\sin \alpha \sin \beta}.$$

For the regular icosahedron, $\alpha = 60°$, $\beta = 60°$, and $\gamma = 108°$ (the angle between two adjacent sides of a regular pentagon).

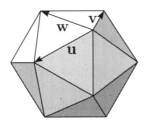

FIGURE 29

Substituting these angles into the last formula, and using $\cos 108° = \frac{1 - \sqrt{5}}{4}$, we find $\theta = \cos^{-1}\left(-\sqrt{5}/3\right) \approx 138.2°$.

Problem 62

Given a constant C, find all functions f such that

$$f(x) + C f(2 - x) = (x - 1)^3$$

for all x.

Solution. For $C \neq 1, -1$, the unique solution is given by

$$f(x) = \frac{(x - 1)^3}{1 - C};$$

for $C = 1$, there is no solution; for $C = -1$, the solutions are the functions of the form $f(x) = \frac{1}{2}(x-1)^3 + E(x-1)$, where E is any even function.

Replace x by $2 - x$ in the original equation, to get

$$f(2-x) + Cf(x) = -(x-1)^3.$$

We now have a system of two linear equations in the two unknowns $f(x)$ and $f(2-x)$:

$$\begin{pmatrix} 1 & C \\ C & 1 \end{pmatrix} \begin{pmatrix} f(x) \\ f(2-x) \end{pmatrix} = \begin{pmatrix} (x-1)^3 \\ -(x-1)^3 \end{pmatrix}.$$

The determinant of the matrix of coefficients is $1 - C^2$. Thus, if $C^2 \neq 1$, there is a unique solution, namely

$$f(x) = \frac{(x-1)^3}{1-C}.$$

If $C = 1$, then there is no solution, because then for any $x \neq 1$, the equations are inconsistent.

Finally, if $C = -1$, it is straightforward to show that any function of the form $f(x) = \frac{1}{2}(x-1)^3 + E(x-1)$, where E is any even function, satisfies the equation. Conversely, suppose that f satisfies the equation. Then we can write f in the form

$$\begin{aligned} f(x) &= \tfrac{1}{2}\big(f(x) - f(2-x)\big) + \tfrac{1}{2}\big(f(x) + f(2-x)\big) \\ &= \tfrac{1}{2}(x-1)^3 + \tfrac{1}{2}\big(f(x) + f(2-x)\big) \\ &= \tfrac{1}{2}(x-1)^3 + \tfrac{1}{2}\Big(f\big(1+(x-1)\big) + f\big(1-(x-1)\big)\Big) \\ &= \tfrac{1}{2}(x-1)^3 + E(x-1), \end{aligned}$$

where the even function E is defined by $E(x) = \frac{1}{2}\big(f(1+x) + f(1-x)\big)$.

Comment. Another way to solve $f(x) - f(2-x) = (x-1)^3$ is to observe that it is an inhomogeneous linear equation in f, which we could try to solve by first solving the corresponding homogeneous equation $f(x) - f(2-x) = 0$ and then finding a particular solution to the inhomogeneous equation. It is not hard to guess a particular solution, $\frac{1}{2}(x-1)^3$, and it is easy to see that the homogeneous equation is satisfied exactly when f is an even function in $x-1$. Thus, the general solution is $f(x) = \frac{1}{2}(x-1)^3 + E(x-1)$, where E is an even function.

Problem 63

A digital time/temperature display flashes back and forth between time, temperature in degrees Fahrenheit, and temperature in degrees Centigrade. Suppose that over the course of a week in spring, the temperatures measured are between 15°C and 25°C and that they are randomly and uniformly distributed over that interval. What is the probability that, at any given time, the rounded value in degrees F of the converted temperature (from degrees C) is not the same as the value obtained by first rounding the temperature in degrees C, then converting to degrees F and rounding once more?

Idea. A good way to begin this problem is to consider the step functions which represent the two methods of converting.

Solution 1. The probability that the display will seem to be in error as explained in the problem is 4/9.

To see this, first note that since 5°C exactly equals 9°F, we can restrict ourselves to the interval from 15°C to 20°C. (Whatever happens there will exactly repeat from 20°C to 25°C, except that all readings—rounded or unrounded—will be 5°C, or 9°F, higher.) If we round a temperature in this interval to the nearest degree C and then convert to degrees F, we will get one of the following: 59°F (= 15°C), 60.8°F (= 16°C), 62.6°F, 64.4°F, 66.2°F, 68°F (= 20°C). These round to 59, 61, 63, 64, 66, 68 degrees F respectively. Thus we have the following:

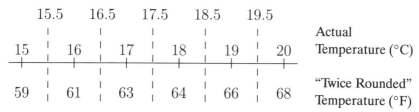

FIGURE 30

On the other hand, if we first convert the temperature in degrees C to degrees F and then round off, we can get any value from 59 to 68, as shown in Figure 31. Comparing the two "thermometers", we see that the "once rounded" and "twice rounded" readings in degrees F are different when the"once rounded" temperature is 60, 62, 65, or 67°F, but the same in all other cases. So out of a total temperature interval of 5°C, there are four intervals, each of length $10/18 = 5/9°$C, in which the display will appear to be in error; thus the probability of this occurring is $(4 \cdot 5/9)/5 = 4/9$, as stated.

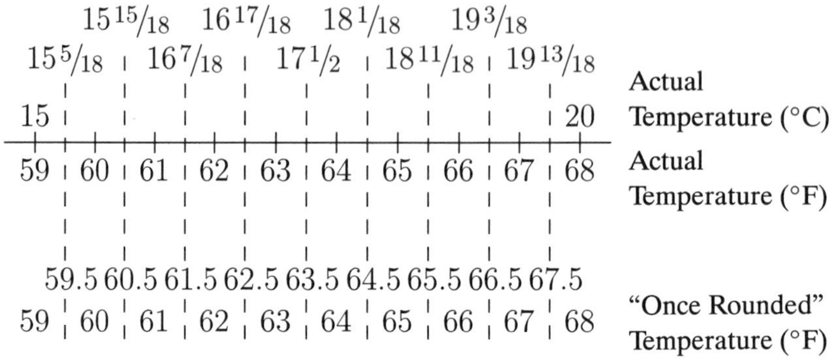

$$15\,^{15}/_{18} \quad 16\,^{17}/_{18} \quad 18\,^{1}/_{18} \qquad 19\,^{3}/_{18}$$
$$15\,^{5}/_{18} \quad \mid \quad 16\,^{7}/_{18} \quad \mid \quad 17\,^{1}/_{2} \quad \mid \quad 18\,^{11}/_{18} \quad \mid \quad 19\,^{13}/_{18}$$

Actual
Temperature (°C)

$$15 \mid \qquad \mid \qquad \mid \qquad \mid \qquad \mid \qquad \mid \qquad \mid \qquad \mid \qquad \mid \; 20$$

Actual
Temperature (°F)

$$59 \mid 60 \mid 61 \mid 62 \mid 63 \mid 64 \mid 65 \mid 66 \mid 67 \mid 68$$

$$59.5\;60.5\;61.5\;62.5\;63.5\;64.5\;65.5\;66.5\;67.5$$
$$59 \mid 60 \mid 61 \mid 62 \mid 63 \mid 64 \mid 65 \mid 66 \mid 67 \mid 68$$

"Once Rounded"
Temperature (°F)

FIGURE 31

Solution 2. Let x be the exact Centigrade temperature. If $f(x)$ denotes the result of converting to Fahrenheit and then rounding, we have

$$f(x) = \lfloor 1.8x + 32.5 \rfloor.$$

This function is a step function with discontinuities $5/9$ apart. On the other hand, if we round, and then convert to Fahrenheit, and round again, we obtain the function

$$F(x) = \lfloor 1.8\lfloor x + .5 \rfloor + 32.5 \rfloor,$$

which has discontinuities at the half-integers.

If n is an integer, we observe that

$$f(n - 1/18) = F(n - 1/18) = \lfloor 1.8n + 32.5 \rfloor,$$

and that

$$\lim_{x \to (n+1/18)^-} f(x) = \lim_{x \to (n+1/18)^-} F(x) = \lfloor 1.8n + 32.5 \rfloor.$$

Thus, f and F agree on the interval $[n - 1/18, n + 1/18)$, which is the middle ninth of a step for $F(x)$; therefore this step must include exactly one entire step of f.

Since $f(x+5) = f(x) + 9$ and $F(x+5) = F(x) + 9$, the probability of agreement is the same on any interval of length 5°C. Since the interval from 15°C to 25°C is made up of two intervals of length 5°C, we may, equivalently, compute the probability on the interval from $-.5$°C to 4.5°C. This interval is comprised of exactly five full steps of $F(x)$, hence the probability of agreement is 5/9 and the probability of discrepancy is 4/9.

Comment. The probability of disagreement if one converts from Fahrenheit to Centigrade over an interval of 9°F is drastically different; similar arguments to those above show that the probability of disagreement is only 2/15. This is due to the fact that the Fahrenheit scale provides a finer measurement.

Problem 64

Sketch the set of points (x, y) in the plane which satisfy

$$(x^2 - y^2)^{2/3} + (2xy)^{2/3} = (x^2 + y^2)^{1/3}.$$

Idea. The presence of sums and differences of squares suggests changing the equation into polar form.

Solution.

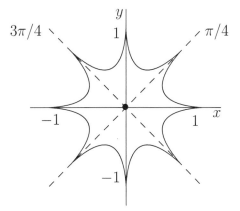

FIGURE 32

To find this graph (which includes an isolated point at the origin), start by transforming the given equation to polar coordinates (using $x = r \cos \theta$, $y = r \sin \theta$, $x^2 + y^2 = r^2$). We get

$$\left(r^2(\cos^2 \theta - \sin^2 \theta)\right)^{2/3} + \left(2r^2 \cos \theta \sin \theta\right)^{2/3} = (r^2)^{1/3}.$$

This is certainly true at the origin (where $r = 0$). For $r \neq 0$, we can divide through by $r^{2/3}$ to obtain

$$\left(r(\cos^2 \theta - \sin^2 \theta)\right)^{2/3} + \left(2r \cos \theta \sin \theta\right)^{2/3} = 1,$$

or

$$(r \cos 2\theta)^{2/3} + (r \sin 2\theta)^{2/3} = 1.$$

If we introduce new "rectangular" coordinates (X, Y) by $X = r \cos 2\theta$, $Y = r \sin 2\theta$, then the last equation reads $X^{2/3} + Y^{2/3} = 1$, which is the equation of an astroid in the XY-plane. Its graph may be found using standard calculus techniques and is shown in Figure 33.

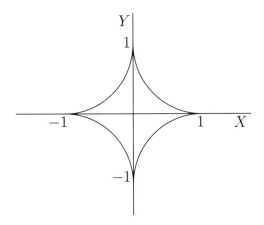

FIGURE 33

Now note that if (x, y) has polar coordinates (r, θ), then the corresponding (X, Y) has polar coordinates $(r, 2\theta)$: the same distance to the origin, but twice the polar angle. Thus we can find the actual graph (in the xy-plane) by replacing each point of the astroid by one with *half* its polar angle, and including the origin. The resulting curve will have cusps at $\theta = 0, \pi/4, \pi/2, 3\pi/4, \ldots$ to correspond to $2\theta = 0, \pi/2, \pi, 3\pi/2, \ldots$, and we get the eight-pointed "star" shown in Figure 32.

Problem 65

Find all integer solutions to $x^2 + 615 = 2^n$.

Idea. If n were even, then $615 = 2^n - x^2$ would be a difference of squares and could be factored accordingly. Is there any reason for n to be even?

Solution. The only solutions are $(x, n) = (\pm 59, 12)$.

We first factor 615 as $3 \cdot 5 \cdot 41$. Since 2 is not a square modulo 3 (or mod 5), n must be even, say $n = 2m$. Then $615 = 2^{2m} - x^2 = (2^m - x)(2^m + x)$. Now 615 can be written as a product of two positive integers in four ways: $1 \cdot 615$, $3 \cdot 205$, $5 \cdot 123$, and $15 \cdot 41$. Since $(2^m - x) + (2^m + x) = 2^{m+1}$, we want the sum of the two factors to be a power of 2. This occurs only for the factorization $615 = 5 \cdot 123$, which yields the solutions $x = \pm 59$, $n = 12$.

Problem 66

Sum the infinite series

$$\sum_{n=1}^{\infty} \sin \frac{2\alpha}{3^n} \sin \frac{\alpha}{3^n}.$$

Idea. The best hope of computing the partial sums is to show that they telescope.

Solution. The sum is $\dfrac{1 - \cos \alpha}{2}$.

Using the identity $\sin \theta \sin \phi = \frac{1}{2} \cos (\theta - \phi) - \frac{1}{2} \cos (\theta + \phi)$, we find that the kth partial sum is

$$S_k = \sum_{n=1}^{k} \sin \left(\frac{2\alpha}{3^n} \right) \sin \left(\frac{\alpha}{3^n} \right)$$

$$= \sum_{n=1}^{k} \left(\frac{1}{2} \cos \left(\frac{\alpha}{3^n} \right) - \frac{1}{2} \cos \left(\frac{\alpha}{3^{n-1}} \right) \right)$$

$$= \frac{1}{2} \cos \left(\frac{\alpha}{3^k} \right) - \frac{1}{2} \cos \alpha.$$

It follows that

$$\sum_{n=1}^{\infty} \sin\left(\frac{2\alpha}{3^n}\right) \sin\left(\frac{\alpha}{3^n}\right) = \lim_{k \to \infty} \left(\frac{1}{2}\cos\left(\frac{\alpha}{3^k}\right) - \frac{1}{2}\cos\alpha\right) = \frac{1-\cos\alpha}{2}.$$

Problem 67

Do there exist five rays emanating from the origin in $\mathbf{R}^3$ such that the angle between any two of these rays is obtuse?

Solution. No, not all the angles formed by pairs of the five rays can be obtuse. Let $v_i = (x_i, y_i, z_i)$, $1 \le i \le 5$, denote a unit vector along the ith ray, and suppose that all the angles are obtuse. By proper choice of axes (corresponding to a rotation), we may assume $v_5 = (0, 0, -1)$. We have

$$\mathbf{v}_i \cdot \mathbf{v}_j = |\mathbf{v}_i||\mathbf{v}_j| \cos\theta_{ij} < 0,$$

where θ_{ij} is the (obtuse) angle between v_i and v_j. In particular, taking $j = 5$ we see that $z_i > 0$ for $i = 1, 2, 3, 4$.

Now let w_i be the projection of v_i onto the xy-plane, i.e., $\mathbf{w}_i = (x_i, y_i, 0)$. Some pair of w_1, w_2, w_3, w_4 form a non-obtuse angle. We may assume that w_1, w_2 is such a pair, so that $0 \le \mathbf{w}_1 \cdot \mathbf{w}_2 = x_1 x_2 + y_1 y_2$. But then

$$\mathbf{v}_1 \cdot \mathbf{v}_2 = x_1 x_2 + y_1 y_2 + z_1 z_2 > 0,$$

and the angle between v_1 and v_2 is acute, completing the proof.

Comment. No two of the *six* unit vectors $(\pm 1, 0, 0)$, $(0, \pm 1, 0)$, $(0, 0, \pm 1)$ along the coordinate axes form an *acute* angle.

Problem 68

Find all twice continuously differentiable functions f for which there exists a constant c such that, for all real numbers a and b,

$$\left| \int_a^b f(x)\, dx - \frac{b-a}{2}\left(f(b) + f(a)\right) \right| \le c(b-a)^4.$$

Idea. Consider

$$\lim_{b \to a} \frac{\int_a^b f(x)\, dx - \frac{b-a}{2}\left(f(b) + f(a)\right)}{(b-a)^3}.$$

Solution. The functions f are the linear polynomials.

First of all, if f is a linear polynomial, then

$$\int_a^b f(x)\, dx = \frac{b-a}{2}\left(f(b) + f(a)\right).$$

(This can be shown either by direct computation, or by interpreting the right-hand side as the signed area of the trapezoid bounded by $x = a, x = b$, the x-axis, and $y = f(x)$.) Thus if f is a linear polynomial, then the inequality in the problem is satisfied for $c = 0$.

Conversely, suppose that f is a twice continuously differentiable function which satisfies the given inequality for some c. For fixed a, we can consider

$$L = \lim_{b \to a} \frac{\int_a^b f(x)\, dx - \frac{b-a}{2}\left(f(b) + f(a)\right)}{(b-a)^3}.$$

On the one hand, the inequality shows that $L = 0$. On the other hand, L is the limit of an indeterminate form, and l'Hôpital's rule can be applied. By the Fundamental Theorem of Calculus and two applications of l'Hôpital's rule, we have

$$L = \lim_{b \to a} \frac{f(b) - \frac{1}{2}\left(f(b) + f(a)\right) - \frac{b-a}{2}f'(b)}{3(b-a)^2} = \lim_{b \to a} \frac{f(b) - f(a) - (b-a)f'(b)}{6(b-a)^2}$$

$$= \lim_{b \to a} \frac{f'(b) - f'(b) - (b-a)f''(b)}{12(b-a)} = \lim_{b \to a} -\frac{f''(b)}{12} = -\frac{f''(a)}{12},$$

since f'' is continuous. Therefore, $f''(a) = 0$ for all a, that is, f'' is identically zero. So we have $f(x) = C_1 x + C_2$ for some constants C_1 and C_2, and we are done.

Comment. By the same argument, our conclusion still holds if we replace the exponent 4 on the right-hand side of the inequality by $3 + \varepsilon$ for any $\varepsilon > 0$. On the other hand, the error in the trapezoidal estimate $\frac{b-a}{2}\left(f(b) + f(a)\right)$ for the integral $\int_a^b f(x)\, dx$ is known to have the form $-f''(\xi)\frac{(b-a)^3}{12}$ for some ξ between a and b. Thus if the exponent 4 were replaced by 3 instead of $3+\varepsilon$, any function f whose second derivative is bounded and continuous would qualify.

Problem 69

Consider the following two-person game, in which players take turns coloring edges of a cube. Three colors (red, green, and yellow) are available. The cube starts off with all edges uncolored; once an edge is colored, it cannot be colored again. Two edges with a common vertex are not allowed to have the same color. The last player to be able to color an edge wins the game.

a. Given best play on both sides, should the first or the second player win? What is the winning strategy?

b. Since there are twelve edges in all, a game can last at most twelve turns. How many twelve-turn end positions are essentially different?

Solution.   a.   The second player should win. A winning strategy for the second player is to duplicate each of the first player's moves on the opposite side of the cube. That is, whenever the first player colors an edge, the second player counters by coloring the diametrically opposite edge the same color. After each of the second player's moves, the partial coloring of the cube will be symmetric with respect to the center of the cube. Therefore, any new move the first player makes which is legal (i.e., which does not cause two edges of the same color to meet) can be duplicated on the diametrically opposite side by the second player, and so the second player will be the last to be able to move and thus wins the game.

b.   There are four essentially different end positions for a twelve-turn game, as shown below.

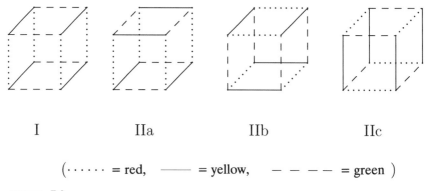

| I | IIa | IIb | IIc |

$(\cdots\cdots = \text{red}, \quad \text{------} = \text{yellow}, \quad \text{-- -- -- --} = \text{green})$

FIGURE **34**

To see why, first note that in an end position for a twelve-turn game, each edge is colored and the three edges that meet at any vertex have the three different colors red, green, yellow, in some order. Now consider the four edges around a face of the cube. Since only three colors are available, two of these edges (at least) must have the same color; these edges must be parallel, or they would meet at a corner of the face. Thus for each face, either one or both pairs of parallel edges of that face must have the same color. Now we can distinguish two cases.

Case 1.   For each face, *both* pairs of parallel edges of that face have the same color. It follows that for each of the three possible directions for edges of a cube, all four edges in that direction have the same color. By rotating the cube, we can assume that the red edges are vertical and the green and yellow edges are as shown in I above. Thus there is only one (essentially different) end position possible in this case.

Case 2.   On at least one face, only one pair of parallel edges has the same color. By rotating the cube, we can assume that this is the front face and that the edges of the same color are vertical. Suppose this color is red. Then by turning the cube upside down if necessary, we can also assume that the top edge of the front face is yellow and the bottom edge is green. But then the colors of the edges leading back are completely determined, and so are the colors of the back edges. Thus the cube is colored as shown in IIa.

Similarly, if the color of the vertical edges in the front face is green, the cube can be turned upside down, if necessary, to match IIb above, whereas if the color is yellow, the cube will match IIc. In each case, all vertical edges have the same color but the edges in each of the other directions are not the same color, so the cases shown in IIa, IIb, and IIc are essentially different.

Thus, there are four essentially different end positions in all.

Problem 70

> Fifty-two is the sum of two squares;
> And three less is a square! So who cares?
> You may think it's curious,
> Perhaps it is spurious,
> Are there other such numbers somewheres?

Are there other solutions in integers? If so, how many?

Answer. There are infinitely many integer solutions.

Solution 1. The only number smaller than 52 with the stated property is 4 (a somewhat degenerate case); the first number larger than 52 with the property is 292. We will show that there are infinitely many integers with the stated property.

Rewrite the equation $x^2 + y^2 = z^2 + 3$ in the form

$$x^2 - 3 = z^2 - y^2 = (z + y)(z - y).$$

If we set $a = z + y$ and $b = z - y$, we have $x^2 - 3 = ab$; also $y = (a - b)/2$ and $z = (a + b)/2$, so y and z will be integers provided a and b have the same parity. This is impossible if x is odd, since $x^2 - 3 \equiv 0 \pmod 4$ has no solution, so let x be even, say $x = 2k$. Then $x^2 - 3 = (4k^2 - 3) \cdot 1$, and therefore we can take $a = 4k^2 - 3$, $b = 1$ and have a and b both be odd. In this way we find the infinite family of solutions

$$\begin{cases} x = 2k \\ y = 2k^2 - 2 \\ z = 2k^2 - 1. \end{cases}$$

Solution 2. Here is an improvement on Solution 1: we will find a doubly infinite family of solutions.

We begin as in Solution 1, but we will now choose x to be a suitable polynomial in k and m, so that $x^2 - 3$ will have a polynomial factorization.

Keeping x even, we can make sure that $x^2 - 3$ has the same polynomial factor $4k^2 - 3$ as in Solution 1 by setting $x = 2m(4k^2 - 3) + 2k$, for then

$$x^2 - 3 = (4k^2 - 3)\big(4m^2(4k^2 - 3) + 8mk + 1\big).$$

Solving

$$\begin{cases} z + y = 4m^2(4k^2 - 3) + 8mk + 1 \\ z - y = 4k^2 - 3, \end{cases}$$

we find

$$\begin{cases} x = 2m(4k^2 - 3) + 2k \\ y = 2m^2(4k^2 - 3) + 4mk - 2k^2 + 2 \\ z = 2m^2(4k^2 - 3) + 4mk + 2k^2 - 1. \end{cases}$$

Problem 71

Starting with a positive number $x_0 = a$, let $(x_n)_{n\geq0}$ be the sequence of numbers such that

$$x_{n+1} = \begin{cases} x_n^2 + 1 & \text{if } n \text{ is even,} \\ \sqrt{x_n} - 1 & \text{if } n \text{ is odd.} \end{cases}$$

For what positive numbers a will there be terms of the sequence arbitrarily close to 0?

Solution. The sequence has terms arbitrarily close to 0 for all positive numbers a. In fact, the subsequence $(x_{2n})_{n\geq0}$ has limit 0.

To show this, let $y_n = x_{2n}$. We have

$$y_{n+1} = \sqrt{x_{2n+1}} - 1 = \sqrt{x_{2n}^2 + 1} - 1 = \sqrt{y_n^2 + 1} - 1, \qquad (*)$$

from which we conclude $y_n > 0$ for all n. Furthermore,

$$y_{n+1} < \sqrt{y_n^2 + 2y_n + 1} - 1 = y_n.$$

We have just seen that (y_n) is a decreasing sequence of positive numbers, and as such it must have a limit L, $L \geq 0$. We finish the solution by proving that $L = 0$. Taking the limit of each side of $(*)$ yields

$$L = \sqrt{L^2 + 1} - 1,$$

or

$$(L + 1)^2 = L^2 + 1.$$

The only solution to this equation is $L = 0$, and we are done.

Problem 72

a. Find all positive numbers T for which

$$\int_0^T x^{-\ln x}\, dx = \int_T^\infty x^{-\ln x}\, dx.$$

b. Evaluate the above integrals for all such T, given that $\int_0^\infty e^{-x^2}\, dx = \sqrt{\pi}/2$.

Solution.   We will show that $T = e^{1/2}$ is the only such number and that

$$\int_0^T x^{-\ln x}\,dx = \int_T^\infty x^{-\ln x}\,dx = \frac{e^{1/4}\sqrt{\pi}}{2}.$$

Substituting $y = \ln x$ converts the integrals

$$\int_0^T x^{-\ln x}\,dx \quad \text{and} \quad \int_T^\infty x^{-\ln x}\,dx$$

to

$$\int_{-\infty}^{\ln T} e^{-y^2+y}\,dy \quad \text{and} \quad \int_{\ln T}^\infty e^{-y^2+y}\,dy.$$

In order to make these integrals more symmetric, we complete the square in the exponent and substitute $z = y - 1/2$, obtaining

$$e^{1/4}\int_{-\infty}^{\ln T-1/2} e^{-z^2}\,dz \quad \text{and} \quad e^{1/4}\int_{\ln T-1/2}^\infty e^{-z^2}\,dz.$$

These improper integrals both converge.

a.   Because e^{-z^2} is a positive, even function, the two integrals are equal if and only if $\ln T - 1/2 = 0$, or equivalently, $T = e^{1/2}$.

b.   Since $\int_0^\infty e^{-x^2}\,dx = \sqrt{\pi}/2$, we find

$$\int_0^{e^{1/2}} x^{-\ln x}\,dx = \int_{e^{1/2}}^\infty x^{-\ln x}\,dx = \frac{e^{1/4}\sqrt{\pi}}{2}.$$

Problem 73

Let $f(x,y) = x^2 + y^2$ and $g(x,y) = x^2 - y^2$. Are there differentiable functions $F(z)$, $G(z)$, and $z = h(x,y)$ such that $f(x,y) = F(z)$ and $g(x,y) = G(z)$?

Solution.   No. If there were such functions, taking partial derivatives would yield

$$2x = \frac{\partial f}{\partial x} = F'(z)\frac{\partial h}{\partial x}, \qquad 2x = \frac{\partial g}{\partial x} = G'(z)\frac{\partial h}{\partial x},$$

$$2y = \frac{\partial f}{\partial y} = F'(z)\frac{\partial h}{\partial y}, \qquad -2y = \frac{\partial g}{\partial y} = G'(z)\frac{\partial h}{\partial y}.$$

We would then have

$$2x\frac{\partial h}{\partial y} = F'(z)\frac{\partial h}{\partial x}\frac{\partial h}{\partial y} = 2y\frac{\partial h}{\partial x}, \qquad 2x\frac{\partial h}{\partial y} = G'(z)\frac{\partial h}{\partial x}\frac{\partial h}{\partial y} = -2y\frac{\partial h}{\partial x},$$

which combine to yield $2x\frac{\partial h}{\partial y} = 2y\frac{\partial h}{\partial x} = 0$. Thus we would have $\frac{\partial h}{\partial y} = \frac{\partial h}{\partial x} = 0$, at least off the x- and y-axes, so h, and therefore f and g, would be constant in the interior of each quadrant, a contradiction.

Problem 74

A point $P = (a, b)$ in the plane is *rational* if both a and b are rational numbers. Find all rational points P such that the distance between P and every rational point on the line $y = 13x$ is a rational number.

Answer.   There are no such points.

Solution 1.   Let the rational point $P = (a, b)$ have rational distance to every rational point on the line $y = 13x$. If P is on the line, then the distance from P to the origin is $\sqrt{170}\,|a|$. Since 170 is not a perfect square, this distance is irrational, unless $a = 0$. However, for $a = 0$, the distance from P to $(1, 13)$ is irrational. Thus, P cannot be on the line.

   Now suppose that P is not on the line. Let $Q = (c, d)$ be the foot of the perpendicular from P to the line.

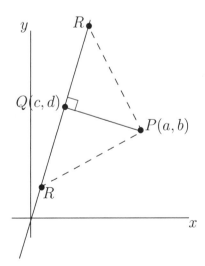

FIGURE 35

Then Q is the intersection of the lines

$$y = 13x \quad \text{and} \quad y = -\frac{1}{13}(x - a) + b,$$

so Q is a rational point. Therefore, the distance PQ is rational. Starting at Q, we move a distance PQ along the line $y = 13x$. The two points R we could end up at are

$$(b + c - d, -a + c + d) \quad \text{and} \quad (-b + c + d, a - c + d),$$

which are both rational. However, the distance PR is $\sqrt{2}$ times the distance PQ, hence it cannot be rational, a contradiction.

Solution 2. We show that, more generally, no *real* point P can satisfy the condition. Suppose $P = (a, b)$ does satisfy the condition. Then for any rational number x, the quantity

$$(x - a)^2 + (13x - b)^2 = 170x^2 - (2a + 26b)x + (a^2 + b^2)$$

is the square of a rational number. In particular, setting $x = 0$, we see that $a^2 + b^2$ is the square of a rational number. Setting $x = 1$ (or any nonzero rational) shows that $2a + 26b$ is rational. If $2a + 26b \neq 0$, then we can set $x = (a^2 + b^2)/(2a + 26b)$ and conclude that $170x^2$ is the square of a rational number, so 170 is a perfect square, a contradiction. On the other hand, if $2a + 26b = 0$, substituting $x = \sqrt{a^2 + b^2}$ implies $171(a^2 + b^2)$ is the square of a rational number. Since $a^2 + b^2$ is also a rational square, this implies $a = b = 0$. But then the distance from $P = (0, 0)$ to $(1, 13)$ is the irrational number $\sqrt{170}$, a contradiction.

Solution 3. Let $P = (a, b)$ be a point with the given property. First we show that if k is any positive rational number, then $kP = (ka, kb)$ also has the property. To see this, note that for any rational point $Q = (t, u)$ on the line $y = 13x$, $Q/k = (t/k, u/k)$ is also a rational point on this line. So the distance $d(kP, Q)$ is rational, since $d(kP, Q) = k\, d(P, Q/k)$; thus, kP has the given property. Therefore, we may assume that a and b are *integers*; all the points $P_n = (na, nb)$ will then be lattice points with the given property. As in the first solution, we know that $P\,(= P_1)$ is not on the line $y = 13x$.

Let $Q = (1, 13)$, $O = (0, 0)$, and $d = d(O, P_1)$. For each positive integer n, let $L_n = d(P_n, Q)$. Because the coordinates of Q and all P_n are integers, the rational numbers d and L_n are in fact *integers*.

Now consider triangle QP_nP_{n+1} (see Figure 36). As n increases, $\angle QP_nP_{n+1}$ approaches $180°$, and $L_{n+1} - L_n$ approaches d. (For a formal proof of this, let

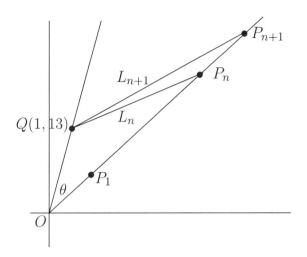

FIGURE 36

$c = d(Q, O)$, and let $\theta = \angle QOP_n$. Then, by the law of cosines,

$$L_n = \sqrt{c^2 + (nd)^2 - 2cnd\cos\theta},$$

and therefore

$$\lim_{n\to\infty}(L_{n+1} - L_n) = \lim_{n\to\infty}\frac{\frac{1}{n}(L_{n+1}^2 - L_n^2)}{\frac{1}{n}(L_{n+1} + L_n)} = \frac{2d^2}{\sqrt{d^2} + \sqrt{d^2}} = d.)$$

Since any convergent sequence of integers must ultimately be constant, it follows that for large n, $L_{n+1} = L_n + d$. On the other hand, the triangle inequality implies that $L_{n+1} < L_n + d$, a contradiction.

Solution 4. As in Solution 2, we will show that there cannot exist any *real* point $P = (a, b)$ whose distance to every rational point on $y = 13x$ is rational.

Suppose P is such a point. Arguing as in Solution 3, we see that for every nonzero rational k, the distance from (ka, kb) to $(1, 13)$ is rational. This clearly rules out $(a, b) = (0, 0)$. Also, the distance from (a, b) to $(0, 0)$ is rational, so the distance from (ka, kb) to $(0, 0)$ can be made to equal any rational $r > 0$ by suitable choice of rational $k > 0$.

Now let r and s be the distances, and θ be the angle, shown in Figure 37. Then whenever $r > 0$ is rational, s is also, because k is. By the law of cosines, r and s are related by

$$s^2 = r^2 - 2r\sqrt{170}\cos\theta + 170.$$

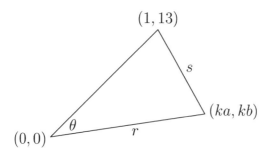

FIGURE 37

The right-hand side of this equation is a polynomial in r of the form

$$r^2 - \alpha r + 170,$$

where $\alpha = 2\sqrt{170}\cos\theta$. Since the value of this polynomial at $r = 1$ is a rational square, α is rational. Note that 170 is divisible by 2 but not by 4. By taking $r > 0$ to be an integer which is divisible by 4 times the denominator of α, we can arrange that $r^2 - \alpha r + 170$ is an integer which is divisible by 2 but not 4, and hence is not a square. But this contradicts the fact that $r^2 - \alpha r + 170$ must be the square of a rational number s.

Problem 75

For what positive numbers x does the series

$$\sum_{n=1}^{\infty}(1 - \sqrt[n]{x}) = (1 - x) + (1 - \sqrt{x}) + (1 - \sqrt[3]{x}) + \cdots$$

converge?

Answer. The series converges only for $x = 1$. Since the convergence for $x = 1$ is obvious, our solutions will assume $x \neq 1$.

Solution 1. We plan to use the limit comparison test, comparing the given series with $\sum_{n=1}^{\infty}\frac{1}{n}$. First we observe that, given x, $1 - \sqrt[n]{x}$ has the same sign for all n. Thus, it is enough to show that $\sum_{n=1}^{\infty}|1 - \sqrt[n]{x}|$ diverges. Writing $\sqrt[n]{x}$

as $e^{(\ln x)/n}$ and applying l'Hôpital's rule, we get

$$\lim_{n\to\infty} \frac{1 - \sqrt[n]{x}}{1/n} = \lim_{n\to\infty} \frac{1 - e^{(\ln x)/n}}{1/n}$$

$$= \lim_{n\to\infty} \frac{-e^{(\ln x)/n} \cdot \left((-\ln x)/n^2\right)}{-1/n^2} = -\ln x.$$

Therefore,

$$\lim_{n\to\infty} \frac{|1 - \sqrt[n]{x}|}{1/n} = |\ln x|,$$

which is a finite nonzero number (since $x \neq 1$). Because $\sum_{n=1}^{\infty} \frac{1}{n}$ diverges, so does $\sum_{n=1}^{\infty} |1 - \sqrt[n]{x}|$.

Solution 2.   If the original series converges, then so does

$$\sum_{n=1}^{\infty} \frac{1 - \sqrt[n]{x}}{1 - x}.$$

Since $1 - x = 1 - (\sqrt[n]{x})^n$, we have

$$\frac{1 - \sqrt[n]{x}}{1 - x} = \frac{1}{1 + \sqrt[n]{x} + (\sqrt[n]{x})^2 + \cdots + (\sqrt[n]{x})^{n-1}}.$$

Suppose first that $x < 1$. Then

$$\frac{1 - \sqrt[n]{x}}{1 - x} > \frac{1}{1 + 1 + 1 + \cdots + 1} = \frac{1}{n},$$

so since $\sum_{n=1}^{\infty} \frac{1}{n}$ diverges, we conclude that

$$\sum_{n=1}^{\infty} (1 - \sqrt[n]{x})$$

also diverges. If, on the other hand, $x > 1$, then

$$\frac{1 - \sqrt[n]{x}}{1 - x} > \frac{1}{(\sqrt[n]{x})^{n-1} + (\sqrt[n]{x})^{n-1} + (\sqrt[n]{x})^{n-1} + \cdots + (\sqrt[n]{x})^{n-1}} > \frac{1}{nx},$$

and since $\sum_{n=1}^{\infty} \frac{1}{nx} = \frac{1}{x} \sum_{n=1}^{\infty} \frac{1}{n}$ diverges, so does $\sum_{n=1}^{\infty} (1 - \sqrt[n]{x})$.

Comments.   The comparison with $\sum_{n=1}^{\infty} \frac{1}{n}$ arises naturally in the second solution, but how could anyone have thought to use it in the first solution? One heuristic approach is to use the binomial theorem for noninteger exponents:

$$(1 + z)^{1/n} = 1 + \frac{\left(\frac{1}{n}\right)}{1!} z + \frac{\left(\frac{1}{n}\right)\left(\frac{1}{n} - 1\right)}{2!} z^2 + \cdots.$$

For $|z|$ small, we have $(1+z)^{1/n} \approx 1+z/n$. Because convergence is most likely for $x \approx 1$, we might start by assuming that

$$1 - \sqrt[n]{x} \approx 1 - (1 + (x-1)/n) = (1-x)/n.$$

The comparison with $\sum_{n=1}^{\infty} \frac{1}{n}$ is then quite natural.

In the first solution, we considered

$$\lim_{n \to \infty} \frac{1 - \sqrt[n]{x}}{1/n}.$$

This limit can also be viewed as the derivative of $f(y) = -x^y = -e^{y \ln x}$ at $y = 0$, which again yields the value $-\ln x$.

Another approach to the problem is to apply the Mean Value Theorem to $f(y)$ to see that $1 - \sqrt[n]{x} = -\frac{1}{n} x^{c_n} \ln x$ for some c_n between 0 and $1/n$. Since for any $x > 0$ and any n, $x^{c_n} \geq \min\{1, x\}$, we see that $|1 - \sqrt[n]{x}|$ is bounded below by a constant (depending on x) times $1/n$.

Problem 76

Let R be a commutative ring with at least one, but only finitely many, (nonzero) zero divisors. Prove that R is finite.

Idea. The key observation is that the product of a zero divisor and an arbitrary element of a commutative ring is either 0 or a zero divisor.

Solution. Let $z_1, z_2, \ldots, z_n$ be the zero divisors in R. Let the set S consist of those nonzero elements of R which are not zero divisors. It is enough to show that S is finite. If $s \in S$, then $z_1 s$ is a zero divisor. Hence $z_1 s = z_j$ for some j. On the other hand, for a fixed j, there are at most $(n+1)$ elements $s \in S$ with $z_1 s = z_j$, because if $z_1 s' = z_1 s$, then $z_1(s' - s) = 0$, hence $s' - s$ is either 0 or a zero divisor. Thus, S has at most $n(n+1)$ elements, and the proof is complete.

Comments. We have actually proved that

$$|R| \leq 1 + n + n(n+1) = (n+1)^2.$$

This inequality is sharp for some n: the ring $\mathbf{Z}_{p^2}$, p prime, has $p-1$ zero divisors.

An analogous result, proved similarly, holds for a noncommutative ring for which the number of left (or right) zero divisors is finite and nonzero.

Problem 77

Let $(a_n)_{n\geq 0}$ be a sequence of positive integers such that $a_{n+1} = 2a_n + 1$. Is there an a_0 such that the sequence consists only of prime numbers?

Idea. Experimenting with several such sequences, we observe that odd primes in the sequence show up as factors of later terms in the sequence, in fact, in a periodic way. This suggests a connection with Fermat's little theorem, which states that if p is a prime and a is not divisible by p, then $a^{p-1} - 1$ is divisible by p.

Solution. There is no such a_0. To see why, first note that $a_{n+1} = 2a_n + 1$ implies $a_{n+1} + 1 = 2(a_n + 1)$ and thus $a_{n+1} + 1 = 2^n(a_1 + 1)$, that is,

$$a_{n+1} = 2^n a_1 + 2^n - 1.$$

Now $a_1 = 2a_0 + 1$ is odd, so if a_1 is also prime, then $2^{a_1-1} - 1$ is divisible by a_1 by Fermat's little theorem. Thus, for $n = a_1 - 1$, $a_{n+1} = 2^n a_1 + (2^{a_1-1} - 1)$ is divisible by a_1, and since $a_{n+1} > a_1 > 1$, a_{n+1} cannot be prime.

Problem 78

Suppose $c > 0$ and $0 < x_1 < x_0 < 1/c$. Suppose also that $x_{n+1} = c x_n x_{n-1}$ for $n = 1, 2, \ldots$.

a. Prove that $\lim_{n\to\infty} x_n = 0$.
b. Let $\phi = (1 + \sqrt{5})/2$. Prove that

$$\lim_{n\to\infty} \frac{x_{n+1}}{x_n^\phi}$$

exists, and find it.

Solution. (Joe Buhler, Reed College)
  a. Define a new sequence by $y_n = c x_n$. Then $0 < y_1 < y_0 < 1$ and $y_{n+1} = y_n y_{n-1}$. An easy induction argument shows that $y_n < y_0^n$ for all $n > 0$, so $y_n \to 0$ and $x_n \to 0$ as $n \to \infty$.

  b. Define

$$z_n = \frac{y_{n+1}}{y_n^\phi} = c^{1-\phi} \frac{x_{n+1}}{x_n^\phi}.$$

We will find $\lim_{n\to\infty} z_n$. Since $y_{n+1} = y_n y_{n-1}$ and $\phi(\phi - 1) = 1$, we have

$$z_n = \frac{y_{n+1}}{y_n^\phi} = \frac{y_{n-1}}{y_n^{\phi-1}} = \left(\frac{y_n}{y_{n-1}^\phi}\right)^{1-\phi} = z_{n-1}^{1-\phi}.$$

By induction, $z_n = z_0^{(1-\phi)^n}$, and since $|1 - \phi| < 1$, we have $\lim_{n\to\infty} z_n = 1$. Therefore,

$$\lim_{n\to\infty} \frac{x_{n+1}}{x_n^\phi} = c^{\phi-1}.$$

Comments. A straightforward induction shows that $y_{n+1} = y_1^{F_{n+1}} y_0^{F_n}$, where F_n denotes the nth Fibonacci number. ($F_0 = 0$, $F_1 = 1$, and $F_n = F_{n-1} + F_{n-2}$ for $n \geq 2$.)

This problem was sparked by the rate of convergence of the secant method in numerical analysis.

Problem 79

Given 64 points in the plane which are positioned so that 2001, but no more, distinct lines can be drawn through pairs of points, prove that at least four of the points are collinear.

Solution. First note that there are

$$\binom{64}{2} = \frac{64 \cdot 63}{2} = 2016$$

ways to choose a pair of points from among the 64 given points. Since there are only 2001 distinct lines through pairs of points, it is clear that some three of the points must be collinear.

Now suppose that no four of the points are collinear. Then every line which connects a pair of the points either contains exactly two or exactly three of the points. If we count lines by counting distinct pairs of points, we will count each line containing exactly two of the points once (which is correct) and each line containing exactly three of the points, say A, B, C, three times (as AB, AC, BC). So to get the correct line count we should subtract 2 for every line containing three of the points. That is, if there are n such lines, then $2001 = 2016 - n \cdot 2$. However, this yields $n = 7\frac{1}{2}$, and n must be an integer, a contradiction.

Comment. One way to create an arrangement satisfying the conditions of the problem is to take four points on each of three lines and 52 other points, while making sure that no other collinearity occurs.

Problem 80

Let $f_1, f_2, \ldots, f_n$ be linearly independent, differentiable functions. Prove that some $n - 1$ of their derivatives $f'_1, f'_2, \ldots, f'_n$ are linearly independent.

Solution 1. If $f'_1, f'_2, \ldots, f'_n$ are linearly dependent, then there is some relation of the form

$$a_1 f'_1 + a_2 f'_2 + \cdots + a_n f'_n = 0,$$

where we may assume $a_n \neq 0$. We will show that $f'_1, f'_2, \ldots, f'_{n-1}$ are then linearly independent. First observe that integrating each side of the preceding equation yields

$$a_1 f_1 + a_2 f_2 + \cdots + a_n f_n = C,$$

where the constant C is nonzero by the linear independence of $f_1, f_2, \ldots, f_n$. If

$$b_1 f'_1 + b_2 f'_2 + \cdots + b_{n-1} f'_{n-1} = 0,$$

then

$$b_1 f_1 + b_2 f_2 + \cdots + b_{n-1} f_{n-1} = D,$$

where D is some constant. We then have

$$(Da_1 - Cb_1)f_1 + (Da_2 - Cb_2)f_2 + \cdots + (Da_{n-1} - Cb_{n-1})f_{n-1} + Da_n f_n = 0.$$

The linear independence of $f_1, f_2, \ldots, f_n$ forces $Da_n = 0$, hence $D = 0$. Therefore, $b_i = 0$ for all i, and $f'_1, f'_2, \ldots, f'_{n-1}$ are linearly independent.

Solution 2. (William C. Waterhouse, Pennsylvania State University) The map taking f to f' is a linear transformation whose kernel (null space) is the 1-dimensional space consisting of the constant functions. The image of any n-dimensional space of differentiable functions must therefore have dimension at least $n - 1$.

Problem 81

Find all real numbers A and B such that

$$\left| \int_1^x \frac{1}{1+t^2} dt - A - \frac{B}{x} \right| < \frac{1}{3x^3}$$

for all $x > 1$.

Answer. The only pair of real numbers which satisfy the given condition is $(A, B) = (\pi/4, -1)$.

If distinct pairs (A, B) and (A', B') satisfied the condition, we would have

$$\left| A - A' + \frac{B - B'}{x} \right| \leq \left| \int_1^x \frac{1}{1+t^2} dt - A - \frac{B}{x} \right| + \left| \int_1^x \frac{1}{1+t^2} dt - A' - \frac{B'}{x} \right|$$

$$< \frac{2}{3x^3}$$

for all $x > 1$, clearly an impossibility. We therefore focus our attention on finding A and B.

Solution 1. We have

$$\int_1^x \frac{1}{1+t^2} dt = \tan^{-1} x - \frac{\pi}{4} = \left(\frac{\pi}{2} - \tan^{-1} \frac{1}{x} \right) - \frac{\pi}{4} = \frac{\pi}{4} - \tan^{-1} \frac{1}{x}.$$

For $|1/x| < 1$, the Taylor series expansion for the inverse tangent implies

$$\tan^{-1} \frac{1}{x} = \frac{1}{x} - \frac{1}{3x^3} + \frac{1}{5x^5} - \frac{1}{7x^7} + \cdots .$$

This series meets the conditions of the alternating series test, hence

$$\left| \tan^{-1} \frac{1}{x} - \frac{1}{x} \right| < \left| \frac{1}{3x^3} \right|.$$

Combining our results, we obtain

$$\left| \int_1^x \frac{1}{1+t^2} dt - \frac{\pi}{4} + \frac{1}{x} \right| < \frac{1}{3x^3}$$

for $x > 1$.

Solution 2. Because $\tan^{-1} t$ is an antiderivative of $1/(1+t^2)$, we have

$$\int_1^\infty \frac{dt}{1+t^2} = \frac{\pi}{4},$$

and therefore

$$\int_1^x \frac{1}{1+t^2}\,dt = \frac{\pi}{4} - \int_x^\infty \frac{1}{1+t^2}\,dt$$

$$= \frac{\pi}{4} - \int_x^\infty \frac{1}{t^2}\,dt - \int_x^\infty \left(\frac{1}{1+t^2} - \frac{1}{t^2}\right)\,dt$$

$$= \frac{\pi}{4} - \frac{1}{x} + \int_x^\infty \frac{1}{(1+t^2)t^2}\,dt.$$

We can estimate this last integral by

$$0 < \int_x^\infty \frac{1}{(1+t^2)t^2}\,dt < \int_x^\infty \frac{1}{t^4}\,dt = \frac{1}{3x^3},$$

and so we have

$$\left|\int_1^x \frac{dt}{1+t^2} - \frac{\pi}{4} + \frac{1}{x}\right| < \frac{1}{3x^3}$$

for all $x > 1$.

Solution 3. As in the second solution, our starting point is

$$\int_1^x \frac{1}{1+t^2}\,dt = \frac{\pi}{4} - \int_x^\infty \frac{1}{1+t^2}\,dt.$$

Converting this latter integral into a series yields

$$\int_x^\infty \frac{1}{1+t^2}\,dt = \int_x^\infty \frac{1}{t^2}\frac{1}{1+1/t^2}\,dt = \int_x^\infty \frac{1}{t^2}\sum_{k=0}^\infty (-1)^k t^{-2k}\,dt.$$

Since $\sum_{k=0}^\infty (-1)^k t^{-2k}$ converges absolutely and uniformly on $[x, \infty)$, we can switch integration and summation to get

$$\int_x^\infty \frac{1}{t^2}\sum_{k=0}^\infty (-1)^k t^{-2k}\,dt = \sum_{k=0}^\infty \frac{(-1)^k}{2k+1}x^{-(2k+1)} = \frac{1}{x} - \frac{1}{3x^3} + \frac{1}{5x^5} - \cdots,$$

and we can proceed as in the first solution.

Problem 82

a. Prove that there is no closed knight's tour of the chessboard which is symmetric under a reflection in one of the main diagonals of the board.

b. Prove that there is no closed knight's tour of the chessboard which is symmetric under a reflection in the horizontal axis through the center of the board.

Solution. a. A closed knight's path, symmetric about a diagonal, can pass through at most two diagonal squares on the axis of reflection. For let A be a diagonal square in the path, and imagine simultaneously traversing the route, in opposite directions, starting at A. The two paths thus traversed are symmetric to each other, so once one of them encounters another diagonal square, so will the other, and the two paths will close to form a closed loop. Since there are eight diagonal squares on the axis of symmetry, there can be no such knight's tour. (In fact, any such tour would require at least four loops.)

  b. Consider a closed path of knight moves that is symmetric about the horizontal axis through the center of the board. Let A denote a square in this path, and let α denote the sequence of moves that goes from A to its mirror image A' (in one of the directions of the closed path). A and A' will be squares of opposite colors (since squares on opposite sides of the horizontal center line have opposite colors), so α consists of an odd number of moves.

None of the individual moves in this "half-loop" from A to A' are symmetric to each other (with respect to the horizontal center line). For if there were such a move, then the moves contiguous (immediately preceding and immediately succeeding) to such a move would also be symmetric, and continuing this argument, the mirror image of each move in α would itself be in α. But this would imply that α is made up of an even number of moves, contradicting the conclusion of the preceding paragraph.

Thus, the path α from A to A' will reflect into a path α' from A' to A. The number of moves in the resulting closed path (α followed by α') is twice an odd number. Conclusion: Any closed path, symmetric about the horizontal axis, must contain twice an odd number of moves. Since 64 is not twice an odd number, no such closed path is possible on the ordinary 8×8 chessboard.

Comments. Similarly, there can be no knight's tour symmetric about a vertical axis through the center.

To complete the analysis of symmetric knight's tours, we sketch a proof that there can be no closed knight's tour which is preserved by a $90°$ rotation about the center. Consider a closed knight's path which is preserved by such a rotation. Pick a square, A, in the path; let A', A'', and A''' be the squares obtained by rotating A through $90°$, $180°$, and $270°$, respectively, about the center. Let α be the sequence of moves from A to A' which does not pass through A''. Since A and A' have opposite color, α consists of an odd number of moves. The result of rotating α through $90°$, say α', connects A' to A''; similarly, α'', α''' connect A'' to A''', A''' to A respectively. The four segments α, α', α'', α''' are disjoint and together form the closed knight's path. Thus the total number of moves is four times an odd number, so it cannot be 64 and the path cannot be a knight's tour. (There *are* closed knight's tours on a 6×6 board which are preserved by a $90°$ rotation; you might enjoy finding one.)

Problem 83

a. Find a sequence (a_n), $a_n > 0$, such that

$$\sum_{n=1}^{\infty} \frac{a_n}{n^3} \quad \text{and} \quad \sum_{n=1}^{\infty} \frac{1}{a_n}$$

both converge.

b. Prove that there is no sequence (a_n), $a_n > 0$, such that

$$\sum_{n=1}^{\infty} \frac{a_n}{n^2} \quad \text{and} \quad \sum_{n=1}^{\infty} \frac{1}{a_n}$$

both converge.

Solution. a. Let $a_n = n^{3/2}$.

b. Suppose there were such a sequence (a_n). Then, since

$$\sum_{n=1}^{\infty} \frac{a_n}{n^2} \quad \text{and} \quad \sum_{n=1}^{\infty} \frac{1}{a_n}$$

would both converge, so would

$$\sum_{n=1}^{\infty} \left(\frac{a_n}{n^2} + \frac{1}{a_n} \right).$$

Now note that

$$\frac{a_n}{n^2} + \frac{1}{a_n} = \left(\frac{\sqrt{a_n}}{n} - \frac{1}{\sqrt{a_n}} \right)^2 + \frac{2}{n} \geq \frac{2}{n},$$

so by the comparison test for positive series, $\sum \frac{2}{n}$ would converge, a contradiction.

Comment. Alternatively, one could observe that if $a_n \geq n$, then

$$\frac{a_n}{n^2} \geq \frac{1}{n},$$

and if $a_n \leq n$, then

$$\frac{1}{a_n} \geq \frac{1}{n}.$$

Hence, for all n,

$$\frac{a_n}{n^2} + \frac{1}{a_n} \geq \max\left\{\frac{a_n}{n^2}, \frac{1}{a_n}\right\} \geq \frac{1}{n}.$$

Also, note that the condition $a_n > 0$ is needed, as shown by the example $a_n = (-1)^n n$.

Problem 84

What is the limit of the repeated power $x^{x^{\cdot^{\cdot^{x}}}}$ with n occurrences of x, as x approaches zero from above?

Answer. The limit is 0 for n odd, 1 for n even.

Solution 1. Obviously, $\lim_{x \to 0+} x = 0$, and it is a standard calculus exercise to show that $\lim_{x \to 0+} x^x = 1$. For $n = 3$, we have $\lim_{x \to 0+} x^{(x^x)} = 0^1 = 0$ and thus we meet the first real problem when $n = 4$.

If we use the standard approach and set $z = x^{(x^{x^x})}$ and proceed to rewrite

$$\ln z = x^{(x^x)} \ln x = \frac{\ln x}{x^{-(x^x)}},$$

we find that l'Hôpital's rule produces a mess. Instead, we look at the size of $\ln z$. Since $\lim_{x \to 0+} x^x = 1$, we know that for (positive) x close enough to 0, $\frac{1}{2} < x^x < 1$. From this we get $x^{1/2} > x^{x^x} > x$ for x close enough to 0. Therefore, $x^{1/2} \ln x < x^{x^x} \ln x < x \ln x$ for x close enough to 0. Now it is not hard to show by l'Hôpital's rule that $\lim_{x \to 0+} x^{1/2} \ln x = \lim_{x \to 0+} x \ln x = 0$. Thus, $\ln z = x^{x^x} \ln x$ is "squeezed" between $x^{1/2} \ln x$ and $x \ln x$, and therefore, $\lim_{x \to 0+} z = e^0 = 1$.

For $n = 5$ we do not have an indeterminate form, and we have

$$\lim_{x \to 0+} x^{\left(x^{x^{x^x}}\right)} = 0^1 = 0.$$

Now that we have discovered the pattern, we show that it will continue by using induction on k to prove that the limit is 0 for $n = 2k - 1$ and 1 for $n = 2k$.

We have seen this for $k = 1$. If it is true for k, then

$$\frac{1}{2} < \underbrace{x^{x^{\cdot^{\cdot^{\cdot^{x}}}}}}_{2k \ x\text{'s}} < 1$$

for x close enough to 0, hence

$$\lim_{x \to 0+} \underbrace{x^{x^{\cdot^{\cdot^{\cdot^{x}}}}}}_{(2k+1) \ x\text{'s}} = 0$$

and even, as in the case $k = 1$ above,

$$\lim_{x \to 0+} \underbrace{x^{x^{\cdot^{\cdot^{\cdot^{x}}}}}}_{(2k+1) \ x\text{'s}} \ln x = 0.$$

This implies

$$\lim_{x \to 0+} \underbrace{x^{x^{\cdot^{\cdot^{\cdot^{x}}}}}}_{(2k+2) \ x\text{'s}} = 1,$$

completing the proof.

Solution 2. We actually prove the following stronger version:

$$\underbrace{x^{x^{\cdot^{\cdot^{\cdot^{x}}}}}}_{n \ x\text{'s}} < x^{1/2} \qquad \text{if } n \text{ is odd and } x \text{ is sufficiently small,}$$

while

$$\lim_{x \to 0+} \underbrace{x^{x^{\cdot^{\cdot^{\cdot^{x}}}}}}_{n \ x\text{'s}} = 1 \qquad \text{if } n \text{ is even.}$$

The proof is by induction on n; the case $n = 1$ is clear. Suppose our claim is true for $n - 1$. If n is odd, then

$$\underbrace{x^{x^{\cdot^{\cdot^{\cdot^{x}}}}}}_{(n-1) \ x\text{'s}} > 1/2$$

for x sufficiently small, and the assertion for n follows. If n is even, then, for x sufficiently small,

$$x^{x^{1/2}} < \underbrace{x^{x^{\cdot^{\cdot^{\cdot^{x}}}}}}_{n \ x\text{'s}} < 1.$$

An easy application of l'Hôpital's rule to $\lim_{x \to 0+} x^{1/2} \ln x$ yields

$$\lim_{x \to 0+} x^{x^{1/2}} = 1,$$

hence

$$\lim_{x \to 0^+} \underbrace{x^{x^{\cdot^{\cdot^{\cdot^{x}}}}}}_{n \ x\text{'s}} = 1$$

as desired. This completes the induction step, and the proof is complete.

Problem 85

A new subdivision is being laid out on the outskirts of Wohascum Center. There are ten north–south streets and six east–west streets, forming blocks which are exactly square. The Town Council has ordered that fire hydrants be installed at some of the intersections, in such a way that no intersection will be more than two "blocks" (really sides of blocks) away from an intersection with a hydrant. (The two blocks need not be in the same direction.) What is the smallest number of hydrants that could be used?

Solution. The smallest number of hydrants required is seven. In fact, seven hydrants suffice even if an additional north–south road is added, as shown by the following map.

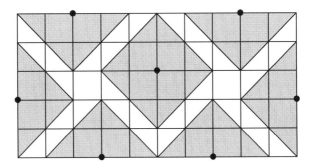

FIGURE 38

To show that six hydrants will not do for the 6×10 grid, consider that there are 28 "edge" vertices (vertices on the outside boundary of the grid) and 12 "center" vertices (vertices 2 "blocks" away from the boundary). Figure 39 indicates how many edge vertices and center vertices are "covered" by hydrants in positions labeled A, B, C.

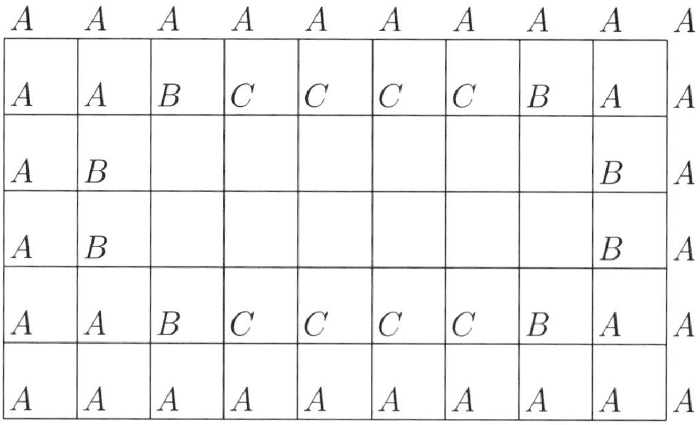

A	A	A	A	A	A	A	A	A	A
A	A	B	C	C	C	C	B	A	A
A	B							B	A
A	B							B	A
A	A	B	C	C	C	C	B	A	A
A	A	A	A	A	A	A	A	A	A

Position of Hydrant	Number of Edge Vertices Covered	Number of Center Vertices Covered
A	5	0 or 1
B	4	3
C	3	4

FIGURE 39

A hydrant placed on a center vertex will cover at most two edge vertices. Therefore, if six hydrants are to cover all 28 edge vertices, their positions must consist of either

(1) 6 of type A,
(2) 5 of type A, 1 of type B,
(3) 5 of type A, 1 of type C, or
(4) 4 of type A, 2 of type B.

However, in case (1) they will cover at most 6 center vertices, and in cases (2), (3), and (4) they will cover at most 8, 9, and 10 center vertices, respectively. Thus, when the edges are covered, the center is not covered. This shows that seven hydrants are necessary.

Problem 86

Consider a triangle ABC whose angles α, β, and γ (at A, B, C respectively) satisfy $\alpha \le \beta \le \gamma$. Under what conditions on α, β, and γ can a beam of light placed at C be aimed at the segment AB, reflect to the segment BC, and then reflect to the vertex A?

Solution.   The necessary and sufficient conditions are $45° - \frac{1}{2}\alpha < \beta < 60°$.

To see why, first recall the reflection principle: If points X, Y are on the same side of a line L, then a light beam originating at X, reflecting in L, and ending at Y, arrives at Y from the direction of X', the reflection of the point X in the line L.

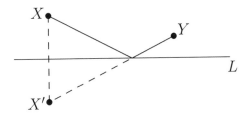

FIGURE 40

In our problem, a light beam placed at C and reflecting to A as described, will arrive at point D on segment BC (after reflecting in AB at E) from the direction of C' and then arrive at A from the direction of C'', where C' is the reflection of C in the line AB and C'' is the reflection of C' in the line BC (see Figure 41).

If AC'' lies within $\angle CAB$, there will be such a light beam, for we can "retrace" the beam from A via D (the intersection point of AC'' and BC) and E (the intersection point of DC' and AB, which exists since α and β are acute) to C. Thus the question becomes: What are the conditions on angles α, β, γ for AC'' to lie within $\angle CAB$?

For AC'' to lie within $\angle CAB$, it is necessary and sufficient that both angles ABC'' and ACC'' shown in Figure 41 be less than $180°$. Now, by symmetry, $\angle CBC'' = \angle CBC' = 2\beta$, so $\angle ABC'' = \angle ABC + \angle CBC'' = 3\beta$, which is less than $180°$ when $\beta < 60°$. On the other hand, since $\triangle CBC''$ is isosceles,

$$\angle BCC'' = \frac{1}{2}(180° - \angle CBC'') = 90° - \beta.$$

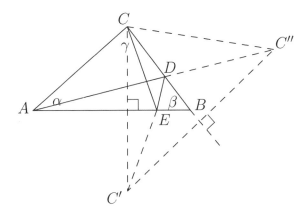

FIGURE 41

Therefore,

$$\angle ACC'' = \angle ACB + \angle BCC'' = \gamma + (90° - \beta) = 270° - \alpha - 2\beta,$$

which is less than $180°$ when $\beta > 45° - \frac{1}{2}\alpha$, and we are done.

Problem 87

Consider the following connect-the-dots game played on an $m \times n$ rectangular lattice. Two players take alternate turns, a turn consisting of drawing a line segment between two points of the lattice so that the interior of the segment intersects neither lattice points nor previously drawn segments. The last player to be able to play wins the game. Which player has the advantage, and what is the winning strategy?

Solution. Assuming optimal play on both sides, the second player will win if m and n are both odd, and the first player will win if at least one of m and n is even. In all cases, the winning strategy is based on the idea of "mirroring" the opponent's moves with respect to the point C at the center of the lattice, and thus restoring symmetry about C at every opportunity.

Suppose one player makes a legal move, drawing segment PQ, when the set of line segments already drawn by both players is symmetric about C. When will it be illegal to respond by drawing $\overline{P}\,\overline{Q}$, where $\overline{P}$ and $\overline{Q}$ are the points opposite P and Q with respect to C? The interior of $\overline{P}\,\overline{Q}$ cannot contain a lattice point, for otherwise the interior of PQ would also and PQ would have been an illegal move. Nor can the interior of $\overline{P}\,\overline{Q}$ intersect a segment RS which was drawn before PQ, because $\overline{R}\,\overline{S}$ would also have been drawn before PQ, and the interior of PQ would have intersected $\overline{R}\,\overline{S}$, making PQ illegal. Thus, $\overline{P}\,\overline{Q}$ is a legal rejoinder unless the interior of $\overline{P}\,\overline{Q}$ intersects PQ. However, if PQ and $\overline{P}\,\overline{Q}$ have any point O in common, they must also have $\overline{O}$ and hence C (which is between O and $\overline{O}$) in common. We can conclude that starting from a symmetric position, $\overline{P}\,\overline{Q}$ will be a legal rejoinder to PQ unless PQ contains C in its interior.

Now suppose m and n are both odd. Then C is itself a lattice point, so PQ can never contain C in its interior. The winning strategy for the second player is to counter each move PQ with its opposite $\overline{P}\,\overline{Q}$.

On the other hand, if either m or n is even, C is not a lattice point, and the first player can win by first drawing a segment $R\overline{R}$ between opposite lattice points, and then restoring symmetry after every move PQ of the second player by drawing $\overline{P}\,\overline{Q}$.

Problem 88

Find

$$\lim_{n \to \infty} \left(\sum_{k=1}^{n} \frac{1}{\binom{n}{k}} \right)^n,$$

or show that this limit does not exist.

Solution.   The limit exists and equals e^2.

To see this, let

$$a_n = \sum_{k=1}^{n} \frac{1}{\binom{n}{k}}.$$

For $n \geq 3$,

$$a_n \geq 1 + \frac{2}{\binom{n}{1}} = 1 + \frac{2}{n},$$

while for $n \geq 6$,

$$a_n = 1 + 2\frac{1}{\binom{n}{1}} + 2\frac{1}{\binom{n}{2}} + \underbrace{\left(\frac{1}{\binom{n}{3}} + \cdots + \frac{1}{\binom{n}{n-3}}\right)}_{n-5 \text{ terms}}$$

$$\leq 1 + \frac{2}{n} + \frac{4}{n(n-1)} + \frac{n-5}{\binom{n}{3}}$$

$$= 1 + \frac{2}{n} + \frac{4}{n(n-1)} + \frac{6(n-5)}{n(n-1)(n-2)}$$

$$< 1 + \frac{2}{n} + \frac{10}{n(n-1)}$$

$$\leq 1 + \frac{2}{n} + \frac{12}{n^2}.$$

But

$$\lim_{n \to \infty} \left(1 + \frac{2}{n}\right)^n = e^2$$

and

$$\lim_{n \to \infty} \left(1 + \frac{2}{n} + \frac{12}{n^2}\right)^n = \lim_{n \to \infty} \left[\left(1 + \frac{2n+12}{n^2}\right)^{n^2/(2n+12)}\right]^{(2n+12)/n} = e^2,$$

so by the squeeze principle, $\lim_{n \to \infty} a_n^n = e^2$.

Problem 89

A person starts at the origin and makes a sequence of moves along the real line, with the kth move being a change of $\pm k$.
a. Prove that the person can reach any integer in this way.
b. If $m(n)$ is the least number of moves required to reach a positive integer n, prove that $\lim_{n \to \infty} m(n)/\sqrt{n}$ exists and evaluate this limit.

Solution. a. A person moving alternately right and left will, after k moves, be at the position given by the kth partial sum of the series

$$1 - 2 + 3 - 4 + 5 - 6 + \cdots.$$

These partial sums are easily seen to be $1, -1, 2, -2, 3, -3, \ldots$; thus every integer can be reached. (If you do not consider 0 reached by its being the starting point, you can write $0 = 1 + 2 - 3$.)

b.   We will show that the limit is $\sqrt{2}$ by finding upper and lower bounds for $m(n)/\sqrt{n}$.

First of all, note that after k moves the maximum possible distance from the origin is $1 + 2 + \cdots + k = k(k + 1)/2$. Thus for $m = m(n)$ we have $m(m+1)/2 \geq n$, or equivalently, $m(m+1) \geq 2n$. It follows that $(m+1)^2 > 2n$, so

$$m + 1 > \sqrt{2n} \qquad \text{and} \qquad \frac{m(n)}{\sqrt{n}} > \sqrt{2} - \frac{1}{\sqrt{n}}.$$

We will now show that

$$\frac{m(n)}{\sqrt{n}} < \sqrt{2} + \frac{3}{\sqrt{n}}.$$

Since

$$\lim_{n \to \infty} \left(\sqrt{2} - \frac{1}{\sqrt{n}} \right) = \lim_{n \to \infty} \left(\sqrt{2} + \frac{3}{\sqrt{n}} \right) = \sqrt{2},$$

we will then be done by the squeeze principle.

The quickest way to get at least as far as n is to move to the right until one reaches or passes position n. As above, this will happen once $k(k + 1)/2 \geq n$, which is certainly true if $k^2/2 \geq n$, or $k \geq \sqrt{2n}$. Even though $\sqrt{2n}$ may not be an integer, $k(k+1)/2 \geq n$ will first happen for some $k = k_0$ with $k_0 < \sqrt{2n}+1$. At this point we will have overshot n by somewhere between 0 and $k_0 - 1$, that is, $1 + 2 + \cdots + k_0 = n + r$ with $0 \leq r \leq k_0 - 1$. If $r = 0$, we have reached n. If $r = 2s(s > 0)$ is even, we can go back and change the sth move from s to $-s$. Since this changes the outcome by $-2s = -r$, we now have a way to reach n in k_0 moves. If $r = 2s + 1$ is odd, we can reach n in $k_0 + 2$ moves by again changing the sth move but also adding two moves at the end:

$$1 + 2 + \cdots + (s - 1) - s + (s + 1) + \cdots + k_0 + (k_0 + 1) - (k_0 + 2)$$

$$= 1 + 2 + \cdots + k_0 - 2s - 1$$

$$= n.$$

In each case, we can reach n in at most $k_0 + 2 < \sqrt{2n} + 3$ moves, so

$$\frac{m(n)}{\sqrt{n}} < \sqrt{2} + \frac{3}{\sqrt{n}}$$

and we are done.

Problem 90

Let $N > 1$ be a positive integer and consider those functions $\varepsilon \colon \mathbf{Z} \to \{1, -1\}$ having period N. For what N does there exist an infinite series $\sum_{n=1}^{\infty} a_n$ with the following properties: $\sum_{n=1}^{\infty} a_n$ diverges, whereas $\sum_{n=1}^{\infty} \varepsilon(n)a_n$ converges for all nonconstant ε (of period N)?

Solution. Series with the given properties exist only for $N = 2$. One example for $N = 2$ is given by $a_n = \frac{1}{n}$. To show that there are no such series for $N > 2$, fix N and assume that $\sum_{n=1}^{\infty} a_n$ is such a series. Define the periodic functions $\varepsilon_j \colon \mathbf{Z} \to \{1, -1\}$, $j = 1, 2, \ldots, N$ by $\varepsilon_j = -1$ if and only if $n \equiv j \pmod{N}$. Then since $\sum_{n=1}^{\infty} \varepsilon_j(n)a_n$ converges for all j,

$$\sum_{j=1}^{N} \sum_{n=1}^{\infty} \varepsilon_j(n)a_n = \sum_{n=1}^{\infty} \left(\sum_{j=1}^{N} \varepsilon_j(n) \right) a_n = \sum_{n=1}^{\infty} (N - 2)a_n$$

converges as well, so $\sum_{n=1}^{\infty} a_n$ converges.

Problem 91

Suppose f is a continuous, increasing, bounded, real-valued function, defined on $[0, \infty)$, such that $f(0) = 0$ and $f'(0)$ exists. Show that there exists $b > 0$ for which the volume obtained by rotating the area under f from 0 to b about the x-axis is half that of the cylinder obtained by rotating $y = f(b)$, $0 \le x \le b$, about the x-axis.

Idea. For b near 0, the differentiability condition ensures that the solid is "morally" a cone, having volume $\frac{1}{3}\pi r^2 h$. For b approaching ∞, the boundedness condition ensures that the solid is "morally" a cylinder, with volume $\pi r^2 h$. Somewhere in between, the coefficient will be $1/2$.

Solution. If we rotate $y = f(b)$, $0 \le x \le b$, about the x- axis, the cylinder we get has volume $\pi b (f(b))^2$. On the other hand, the volume obtained by rotating the area under f from 0 to b is $\int_0^b \pi (f(x))^2 \, dx$. If we define

$$g(b) = \frac{\int_0^b \pi (f(x))^2 \, dx}{\pi b (f(b))^2} = \frac{\int_0^b (f(x))^2 \, dx}{b (f(b))^2},$$

then we need to prove the existence of a positive b for which $g(b) = 1/2$.

We first prove the existence of a b with $g(b) > 1/2$. Since f is increasing and bounded, we must have $\lim_{x \to \infty} f(x) = c$ for some nonnegative constant c. Given ε satisfying $0 < \varepsilon < \min\{c, 1\}$, choose x_0 such that $f(x_0) \geq c - \varepsilon$. Then

$$\int_0^{x_0/\varepsilon} (f(x))^2 \, dx \geq \int_{x_0}^{x_0/\varepsilon} (f(x))^2 \, dx \geq x_0 \left(\frac{1 - \varepsilon}{\varepsilon} \right) (c - \varepsilon)^2,$$

and thus for $b = x_0/\varepsilon$, we have

$$g(b) \geq \frac{(1 - \varepsilon)(c - \varepsilon)^2}{c^2}.$$

Because ε may be arbitrarily small, $g(b)$ can be made arbitrarily close to 1, and in particular, greater than $1/2$.

To show that g takes on values less than $1/2$, we consider two cases. First, suppose that $f'(0) > 0$. Given $\varepsilon > 0$, there exists a $\delta > 0$ for which $0 \leq x < \delta$ implies $|f(x)/x - f'(0)| < \varepsilon$. If, moreover, $\varepsilon < f'(0)$ and $b < \delta$, we then have

$$g(b) \leq \frac{\int_0^b (f'(0) + \varepsilon)^2 x^2 \, dx}{b (f'(0) - \varepsilon)^2 b^2} = \frac{1}{3} \frac{(f'(0) + \varepsilon)^2}{(f'(0) - \varepsilon)^2}.$$

We conclude that g can be arbitrarily close to $1/3$, hence takes on values less than $1/2$.

If $f'(0) = 0$, define h on $[0, \infty)$ by $h(x) = f(x)/x$ for $x > 0$ and $h(0) = 0$. Observe that h is continuous, so h attains a maximum on $[0, 1]$, say at $x = b$. We then have $b > 0$, and

$$g(b) \leq \frac{\int_0^b \left(\frac{f(b)}{b} x \right)^2 dx}{b (f(b))^2} = \frac{1}{3},$$

as desired.

We apply the Intermediate Value Theorem to conclude that there exists a b for which $g(b) = 1/2$.

Comments. The first part of our argument shows that $\lim_{b \to \infty} g(b) = 1$.

If we had assumed that f is continuously differentiable on $[0, \infty)$ and that $f'(0) > 0$, then we could have used l'Hôpital's rule and the Fundamental Theorem of Calculus to show that $\lim_{b \to 0+} g(b) = 1/3$.

Problem 92

Does the Maclaurin series for $e^{x - x^3}$ have any zero coefficients?

Answer. The Maclaurin series has no zero coefficients.

Solution 1. Let $f(x) = e^{x-x^3}$. The product and chain rules imply that $f^{(n)}(x) = g_n(x)e^{x-x^3}$ where $g_n(x)$ is a polynomial with integer coefficients. A few calculations may lead one to conjecture that $g_n(0)$ is always odd, which would imply the assertion. We prove a stronger statement by induction, namely that $g_n(0)$ is odd and $g_n'(0)$ is even. This is clear for $n = 0$, since $g_0(x) = 1$. Now assume it holds for n. We observe that

$$f^{(n+1)}(x) = g_n'(x)e^{x-x^3} + g_n(x)(1 - 3x^2)e^{x-x^3}.$$

Therefore, $g_{n+1}(x) = g_n'(x) + (1 - 3x^2)g_n(x)$, so $g_{n+1}(0) = g_n'(0) + g_n(0)$ is odd. Differentiating $g_{n+1}(x)$ yields

$$g_{n+1}'(x) = g_n''(x) - 6xg_n(x) + (1 - 3x^2)g_n'(x).$$

Since the second derivative of x^2 is 2, the constant term $g_n''(0)$ of $g_n''(x)$ will be even. Thus, $g_{n+1}'(0) = g_n''(0) + g_n'(0)$ is even, and our proof by induction is complete.

Solution 2. (Bruce Reznick, University of Illinois) Using the power series expansion of the exponential, we have

$$e^{x-x^3} = e^x e^{-x^3} = \sum_{i=0}^{\infty} \frac{x^i}{i!} \sum_{j=0}^{\infty} \frac{(-1)^j x^{3j}}{j!}.$$

Multiplying the series on the right and collecting terms, we see that the coefficient of x^n in the Maclaurin series is given by

$$a_n = \sum_{\substack{i+3j=n \\ i\geq 0, j\geq 0}} \frac{(-1)^j}{i!j!}.$$

Now multiply each side by $(n-1)!$ to get

$$(n-1)!a_n = \sum_{\substack{i+3j=n \\ i\geq 0, j\geq 0}} \frac{(-1)^j (n-1)!}{i!j!}$$

$$= \frac{1}{n} + \sum_{\substack{i+3j=n \\ i\geq 0, j>0}} (-1)^j (n-1)\cdots(i+j+1)\binom{i+j}{i}.$$

This shows that $(n-1)!a_n$ is an integer plus $1/n$, so $a_n \neq 0$ for $n \geq 2$; it is easily seen that $a_0 = a_1 = 1$, so a_n is nonzero for all n.

Problem 93

Let a and d be relatively prime positive integers, and consider the sequence $a, a + d, a + 4d, a + 9d, \ldots, a + n^2d, \ldots$. Given a positive integer b, can one always find an integer in the sequence which is relatively prime to b?

Idea. We want $a + n^2d$ to differ from $a \pmod{p}$ for those primes p which divide both a and b, but not for those which divide b but not a.

Solution. Yes, one can always find a positive integer n such that $a + n^2d$ is relatively prime to b. In fact, let n be the product of those prime factors of b which do not divide a. Then if a prime p divides b and not a, p will divide n^2d and hence not divide $a + n^2d$.

If a prime p divides both b and a, then p will not divide n, and since a, d are relatively prime, p will not divide d. Thus p will not divide n^2d and hence not divide $a + n^2d$. It follows that no prime which divides b can divide $a + n^2d$, and we are done.

Problem 94

Find the smallest possible n for which there exist integers $x_1, x_2, \ldots, x_n$ such that each integer between 1000 and 2000 (inclusive) can be written as the sum, without repetition, of one or more of the integers $x_1, x_2, \ldots, x_n$.

Idea. Since a set of 9 elements has 2^9 subsets, including the empty set, 9 integers can form at most $2^9 - 1 = 511$ positive sums, hence there certainly cannot be 1001 distinct, positive sums. On the other hand, it is not hard to construct sets of eleven integers which work, so we must try to rule out sets of 10 integers.

Solution. The smallest possible n is 11. By binary expansion, each integer between 1000 and 2000 can be written as the sum of some of the eleven integers $1, 2, 2^2, \ldots, 2^{10}$.

Now assume that ten integers $x_1, x_2, \ldots, x_{10}$ have the desired property; we will derive a contradiction. First suppose that two of the integers, say x_9 and x_{10}, are greater than 500. Consider a sum s of some of $x_1, x_2, \ldots, x_8$. Since $(s + x_9 + x_{10}) - s \geq 1002$, at most three of the four sums

$$s, \ s + x_9, \ s + x_{10}, \ \text{and} \ s + x_9 + x_{10}$$

can be between 1000 and 2000 (inclusive). There are 2^8 sums s, including 0. Therefore, at most $3 \cdot 2^8 = 768$ sums of $x_1, x_2, \ldots, x_{10}$ can be between 1000 and 2000. Thus no two of $x_1, x_2, \ldots, x_{10}$ can be greater than 500; in other words, at least nine of $x_1, x_2, \ldots, x_{10}$ are less than or equal to 500. There are $\binom{9}{2} = 36$ sums of two of these nine integers; none of these sums is greater than 1000, so at least 35 of them are too small or redundant. Therefore, at most $2^{10} - 35 = 989$ of the integers from 1000 to 2000 can be represented as the sum of some of $x_1, x_2, \ldots, x_{10}$, and we are done.

Comment. The general question of the minimum number of integers needed to represent each integer between M and N (inclusive), where $0 < M < N$, seems to be more difficult. It is clear that at least $\log_2(N - M + 2)$ integers are required. On the other hand, the sets $\{M, 1, 2, 2^2, \ldots, 2^{\lfloor \log_2(N-M) \rfloor}\}$ and $\{1, 2, 2^2, \ldots, 2^{\lfloor \log_2 N \rfloor}\}$ both suffice. Therefore, the smallest possible n satisfies the inequalities

$$\log_2(N - M + 2) \le n \le \min\{\lfloor \log_2(N - M) \rfloor + 2, \lfloor \log_2 N \rfloor + 1\}.$$

Given M and N, at most two integers satisfy the above inequalities. However, the case $M = 1000, N = 2000$ of our problem shows that the upper bound can be attained, while $M = 25$, $N = 67$, $x_1 = 9$, $x_2 = 13$, $x_3 = 16$, $x_4 = 17$, $x_5 = 18$, $x_6 = 19$ shows that the smaller integer can be attained.

Problem 95

Define $(x_n)_{n \ge 1}$ by

$$x_1 = 1, \qquad x_{n+1} = \frac{1}{\sqrt{2}} \sqrt{1 - \sqrt{1 - x_n^2}}.$$

a. Show that $\lim_{n \to \infty} x_n$ exists and find this limit.
b. Show that there is a unique number A for which $L = \lim_{n \to \infty} x_n / A^n$ exists as a finite nonzero number. Evaluate L for this value of A.

Idea. Looking at the first few terms leads us to suspect that the sequence is decreasing. The presence of $\sqrt{1 - x_n^2}$ in the recurrence suggests a "trig substitution."

Solution. a. We show that the limit is 0.

Suppose that $0 < x_n \leq 1$, and let $0 < \theta_n \leq \pi/2$ be such that $\sin \theta_n = x_n$. Then

$$x_{n+1} = \frac{1}{\sqrt{2}} \sqrt{1 - \sqrt{1 - \sin^2 \theta_n}} = \sqrt{\frac{1 - \cos \theta_n}{2}} = \sin(\theta_n/2).$$

This shows that $0 < x_{n+1} < x_n$. Moreover, since $\theta_1 = \pi/2$, it shows that $\theta_n = \pi/2^n$. Thus,

$$\lim_{n \to \infty} x_n = \lim_{n \to \infty} \sin\left(\frac{\pi}{2^n}\right) = 0.$$

b. $L = \pi$, for $A = 1/2$.

Suppose

$$\lim_{n \to \infty} \frac{\sin(\pi/2^n)}{A^n}$$

exists and is nonzero. Since

$$\lim_{x \to 0} \frac{\sin x}{x} = 1,$$

we have

$$\lim_{n \to \infty} \frac{\sin(\pi/2^n)}{A^n} = \lim_{n \to \infty} \frac{\sin(\pi/2^n)}{A^n} \lim_{n \to \infty} \frac{\pi/2^n}{\sin(\pi/2^n)}$$

$$= \lim_{n \to \infty} \frac{\sin(\pi/2^n)}{A^n} \frac{\pi/2^n}{\sin(\pi/2^n)}$$

$$= \pi \lim_{n \to \infty} \frac{1}{(2A)^n}.$$

This latter limit exists and is nonzero if and only if $A = 1/2$, in which case we have

$$\lim_{n \to \infty} \frac{\sin(\pi/2^n)}{A^n} = \pi.$$

Problem 96

Consider the line segments in the xy-plane formed by connecting points on the positive x-axis with x an integer to points on the positive y-axis with y an integer. We call a point in the first quadrant an *I-point* if it is the intersection of two such line segments. We call a point an *L-point* if there is a sequence of distinct I-points whose limit is the given point. Prove or disprove: If (x, y) is an L-point, then either x or y (or both) is an integer.

Solution. The given statement is true: If (x, y) is an L-point, then either x or y is an integer.

Suppose $(\overline{x}, \overline{y})$ is a point in the first quadrant with neither $\overline{x}$ nor $\overline{y}$ an integer. We will show that $(\overline{x}, \overline{y})$ is not an L-point. Choose positive integers M and N for which the line between $(M, 0)$ and $(0, N)$ passes above the point $(\overline{x}, \overline{y})$. Let ε be any positive number less than all three of the following: (i) the distance from the above line to $(\overline{x}, \overline{y})$, (ii) the distance from $\overline{x}$ to the closest integer, and (iii) the distance from $\overline{y}$ to the closest integer.

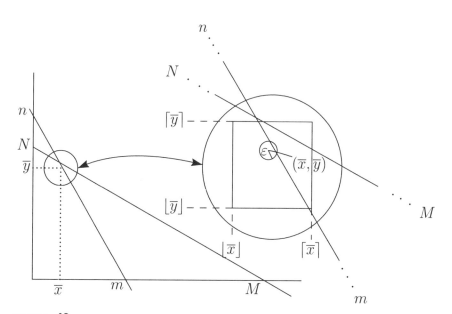

FIGURE 42

We claim that only finitely many lines whose x- and y-intercepts are both positive integers can pass within ε of $(\overline{x}, \overline{y})$. To see this, suppose m and n are positive integers for which the line between $(m, 0)$ and $(0, n)$ passes within ε of $(\overline{x}, \overline{y})$. Then condition (i) above implies that either $m < M$ or $n < N$. Given m, consider the two tangent lines to the circle with center $(\overline{x}, \overline{y})$ and radius ε which pass through $(m, 0)$. Since the line connecting $(m, 0)$ to $(0, n)$ must lie between these tangent lines, we see from condition (ii) above that there are at most finitely many possibilities for n. Similarly, condition (iii) implies that given n, there are at most finitely many possibilities for m. So, indeed, only finitely many lines whose x- and y-intercepts are both positive integers pass within ε of $(\bar{x}, \bar{y})$. Since every I-point within ε of $(\bar{x}, \bar{y})$ has to be one of the finitely many intersections of these lines, the point $(\bar{x}, \bar{y})$ cannot be an L-point, hence the statement is true.

Comment. One can modify the above proof slightly to show that both coordinates of an L-point must be rational.

Problem 97

a. Find all lines which are tangent to both of the parabolas

$$y = x^2 \quad \text{and} \quad y = -x^2 + 4x - 4.$$

b. Now suppose $f(x)$ and $g(x)$ are any two quadratic polynomials. Find geometric criteria that determine the number of lines tangent to both of the parabolas $y = f(x)$ and $y = g(x)$.

Solution. a.   There are two such lines: $y = 0$ and $y = 4x - 4$.
  Suppose the line $y = mx + b$ is tangent to the parabola $y = x^2$ at the point $P = (x_1, x_1^2)$ and is also tangent to the parabola $y = -x^2 + 4x - 4$ at the point $Q = (x_2, -x_2^2 + 4x_2 - 4)$. Then, from calculus, the slope of the line is

$$m = 2x_1 = -2x_2 + 4. \tag{1}$$

First suppose that $x_1 \neq x_2$; then we can also compute the slope from the fact that P and Q are both on the line. This, along with (1), yields

$$m = \frac{-x_2^2 + 4x_2 - 4 - x_1^2}{x_2 - x_1} = \frac{-x_2^2 + 4x_2 - 4 - (2 - x_2)^2}{x_2 - (2 - x_2)} \tag{2}$$

$$= -\frac{(x_2 - 2)^2}{x_2 - 1}.$$

From (1) and (2) we find that $x_2 = 0$ or $x_2 = 2$. For $x_2 = 0$ we get $x_1 = 2$; the tangent line to $y = x^2$ at $(2, 4)$ is $y = 4x - 4$, which is also tangent to $y = -x^2 + 4x - 4$ at $(0, -4)$. For $x_2 = 2$ we get $x_1 = 0$; the tangent line to $y = x^2$ at $(0, 0)$ is $y = 0$, which is also tangent to $y = -x^2 + 4x - 4$ at $(2, 0)$.

  On the other hand, if $x_1 = x_2$, equation (1) implies that $x_1 = x_2 = 1$, $P = (1, 1)$, and $Q = (1, -1)$. That is to say, the line joining P and Q is the vertical line $x = 1$, which is not the tangent line at either point, a contradiction. This completes the consideration of all possibilities.

b.   If $f(x) = g(x)$, there will obviously be infinitely many common tangent lines to $y = f(x)$ and $y = g(x)$. Otherwise, there will be either zero, one, or two common tangent lines. If the two parabolas open in the same direction (i.e., both upward or both downward), then there will be no common tangent line if the parabolas do not intersect, one common tangent line if they intersect in exactly one point, and two common tangent lines if the two parabolas intersect in two distinct points. On the other hand, if the two parabolas open in opposite directions, then there will be two common tangent lines if the parabolas do not intersect, one if they intersect in exactly one point, and none if they intersect in two distinct points. (Note: The parabolas intersect in at most two points since $f(x) - g(x)$ is a nonzero polynomial of degree ≤ 2, so $f(x) - g(x) = 0$ can have at most two solutions.)

To prove these assertions, note that we can shift the coordinate axes so the vertex of $y = f(x)$ is at the origin; also, replacing $f(x)$ by $-f(x)$ and $g(x)$ by $-g(x)$ if necessary, we can assume that the parabola $y = f(x)$ opens upward. Thus, $f(x) = \alpha x^2$ for some $\alpha > 0$. But now, by replacing x by $x/\sqrt{\alpha}$ (a change of scale along the x-axis), we may assume that $f(x) = x^2$. (Of course, $g(x)$ will change accordingly, but it will still be a quadratic polynomial, and the number of common tangent lines and the number of intersection points will not change.)

So we can assume $f(x) = x^2$ and $g(x) = Ax^2 + Bx + C$, $A \neq 0$, $g(x) \neq f(x)$. Let $P = (x_1, x_1^2)$ be a point on $y = f(x)$ and $Q = (x_2, Ax_2^2 + Bx_2 + C)$ be a point on $y = g(x)$. As in (a), if m denotes the slope of a common tangent line, we have

$$m = 2x_1 = 2Ax_2 + B, \tag{1}$$

and if $x_1 \neq x_2$,

$$\begin{aligned} m &= \frac{Ax_2^2 + Bx_2 + C - x_1^2}{x_2 - x_1} \\ &= \frac{(A - A^2)x_2^2 + (1 - A)Bx_2 + (C - B^2/4)}{(1 - A)x_2 - B/2}. \end{aligned} \tag{2}$$

Case 1.   Suppose that $x_1 = x_2$. In this case P and Q must be the same point, because otherwise the line connecting them would be vertical and could not be the tangent line. So we have

$$x_1^2 = Ax_2^2 + Bx_2 + C. \tag{3}$$

Substituting $x_1 = x_2$ into equations (1) and (3) yields

$$(A - 1)x_2 + \frac{B}{2} = 0 \tag{4}$$

and

$$(A - 1)x_2^2 + Bx_2 + C = 0.$$

Subtracting x_2 times (4) from this last equation yields

$$\frac{B}{2}x_2 + C = 0. \tag{5}$$

Equations (4) and (5) will have a simultaneous solution if $B = C = 0$ or if $B \neq 0$ and $(A - 1)C - B^2/4 = 0$. In the first of these cases we see that $g(x) = Ax^2$, $A \neq 1$, and so we find the unique common tangent line $y = 0$, at $(0,0)$. In the second case, we have a common tangent line of slope $-4C/B$ at $(-2C/B, 4C^2/B^2)$.

Case 2. If $x_1 \neq x_2$, combining (1) and (2) yields

$$(A^2 - A)x_2^2 + ABx_2 + \left(\frac{B^2}{4} + C\right) = 0. \tag{6}$$

If $A = 1$, then $B \neq 0$ from (1), since $x_1 \neq x_2$. We find that there is a unique solution, with $x_2 = -B/4 - C/B$ and $x_1 = B/4 - C/B$. If $A \neq 1$, then the quadratic equation (6) for x_2 will have two, one, or no solutions depending on whether its discriminant $\Delta = A\left(B^2 - 4C(A - 1)\right)$ is positive, zero, or negative. Note that if there is a common tangent line under *Case 1*, then either $B = C = 0$ or $B^2 = 4C(A - 1)$; in both cases $\Delta = 0$. If $A \neq 1$ and $\Delta > 0$, there will be exactly two common tangent lines (corresponding to the two solutions for x_2), while if $\Delta < 0$ there will be none. If $A \neq 1$ and $\Delta = 0$, the double root of (6) will be

$$x_2 = -\frac{AB}{2(A^2 - A)} = \frac{B}{2(1 - A)};$$

using (1), this yields

$$x_1 = A\frac{B}{2(1 - A)} + \frac{B}{2} = \frac{B}{2(1 - A)} = x_2,$$

a contradiction. However, in this case the parabolas have exactly one common tangent line, coming from *Case 1*.

To summarize, if $A \neq 1$, the parabolas have two, one, or no common tangent lines depending on whether $\Delta = A\left(B^2 - 4C(A - 1)\right)$ is positive, zero, or negative. If $A = 1$, the parabolas have one common tangent line when $B \neq 0$ and none when $B = 0$.

On the other hand, the intersections of the parabolas are given by

$$x^2 = Ax^2 + Bx + C, \quad \text{or} \quad (A-1)x^2 + Bx + C = 0.$$

If $A \neq 1$, this has two, one, or no solutions according to whether

$$D = B^2 - 4C(A-1)$$

is positive, zero, or negative. Since $\Delta = AD$, we see that if $A > 0$ (parabolas open the same way) and $A \neq 1$, there are two, one, or zero common tangent lines if there are two, one, or zero intersections, while for $A < 0$ (parabolas open in opposite directions) it is the other way around. Finally, if $A = 1$, there is a single intersection point when $B \neq 0$ and no intersection when $B = 0$; the parabolas open the same way, and there are one or zero common tangent lines if there are one or zero intersections, respectively.

Problem 98

Suppose we are given an m-gon and an n-gon in the plane. Consider their intersection; assume this intersection is itself a polygon.

a. If the m-gon and the n-gon are convex, what is the maximal number of sides their intersection can have?
b. Is the result from (a) still correct if only one of the polygons is assumed to be convex?

Solution.   a.   The maximal number of sides that the intersection of a convex m-gon and a convex n-gon can have is $m+n$. To see why more than $m+n$ sides is not possible, first note that any side of the intersection must be part or all of one of the sides of one of the two original polygons, and that there are $m+n$ sides in all of the two original polygons. Thus if the intersection had more than $m+n$ sides, at least two of these would have to be part of the same side of one of the original polygons, which cannot happen because the polygons are convex and hence have convex intersection. Therefore, the intersection can have at most $m+n$ sides.

To show that there really can be $m+n$ sides, we will assume $n \geq m \geq 3$. We will start with a regular $(m+n)$-gon R and show that one can find a convex

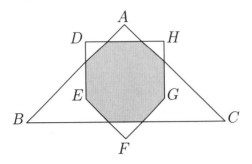

FIGURE 43

m-gon X and a convex n-gon Y whose intersection is R. An example will help illustrate the idea. If $m = 3$ and $n = 5$, then the regular octagon R is the intersection of triangle ABC and pentagon $DEFGH$, as shown.

As in the example, the idea in general is to get the m-gon and the n-gon by extending suitable sides of R; we have to make sure that the sides will intersect when we expect them to and that the resulting polygons are convex. So we want to be able to divide the $m + n$ sides of R into two groups: m of the sides will be parts of sides of X while the other n will be parts of sides of Y. If two adjacent sides of R are both parts of sides of X, the vertex where they meet will be a vertex of X, and similarly for Y. On the other hand, if for two adjacent sides of R one is part of a side of X and the other is part of a side of Y, then both these sides get extended beyond the vertex until they meet the extensions of the "next" sides of X, Y respectively.

To make sure that two successive (parts of) sides of X always meet when they are extended, it is enough to make sure that one is less than halfway around the regular $(m + n)$-gon R from the other; the angle at which these sides meet is then certainly less than 180°. So we will have our convex polygons X and Y by choosing m of the sides of R to be sides of X and the others to be sides of Y, provided we can do this in such a way that two successive sides of X or of Y are never halfway or more around the $(m + n)$-gon from each other.

If $m + n$ is even, such an assignment can be carried out by choosing two opposite sides of R, designating one of them as a side of X and the other as a side of Y, and then designating the sides adjacent to each of the two chosen sides as belonging to the polygon which that chosen side *does not* belong to (see Figure 44). This works since $m \geq 3$, $n \geq 3$; the other sides can be designated at random in such a way as to end up with m sides of X and n sides of Y.

If $m + n$ is odd, there are no "opposite sides" of R, but we can designate a side and the two sides adjacent to the opposite vertex as sides of X, then

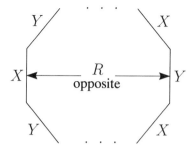

FIGURE 44

designate the four sides adjacent to these three chosen sides as sides of Y (see Figure 45). This works because $m \geq 3$, $n \geq 4$. Once again, all other sides can be designated arbitrarily so as to end up with m sides of X and n sides of Y.

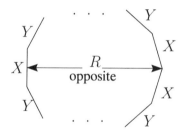

FIGURE 45

b.   No. The following diagram shows that the intersection of a non-convex 4-gon and a convex 3-gon can be an 8-gon.

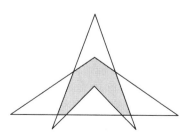

FIGURE 46

Problem 99

If an insect starts at a random point inside a circular plate of radius R and crawls in a straight line in a random direction until it reaches the edge of the plate, what will be the average distance it travels to the edge?

Solution.   The average distance traveled to the edge of the plate is

$$\frac{8}{3\pi} r \approx .8488\, r.$$

We begin by considering the average distance $\overline{D}_X$ to the perimeter if the insect starts at a particular point X inside the circle. Because of symmetry, $\overline{D}_X$ will only depend on the distance from X to the center O of the circle; suppose this distance is a. As shown in the figure, let $f(\alpha)$ be the distance from X to the perimeter along the line at angle α (counterclockwise) from the radius through X.

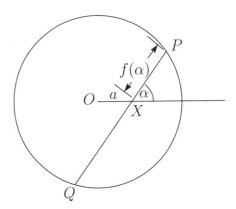

FIGURE 47

Since the insect crawls off in a random direction, we have

$$\overline{D}_X = \frac{1}{2\pi} \int_0^{2\pi} f(\alpha)\, d\alpha$$

as the average distance it will travel from X to the perimeter. Note that the distance from X to the perimeter at Q (in the opposite direction from P) is $f(\alpha + \pi)$, and that

$$f(\alpha) + f(\alpha + \pi) = PQ = 2\sqrt{r^2 - a^2 \sin^2 \alpha}.$$

Thus, we have

$$\overline{D}_X = \frac{1}{2\pi} \int_0^{2\pi} f(\alpha)\,d\alpha = \frac{1}{2\pi} \int_0^{\pi} (f(\alpha) + f(\alpha + \pi))\,d\alpha$$

$$= \frac{1}{\pi} \int_0^{\pi} \sqrt{r^2 - a^2 \sin^2 \alpha}\,d\alpha$$

$$= \frac{2}{\pi} \int_0^{\pi/2} \sqrt{r^2 - a^2 \sin^2 \alpha}\,d\alpha.$$

To get the overall average, we have to average $\overline{D}_X$ over all points X within the circle. This is most easily done using polar coordinates. (Since r is being used for the radius of the circle, we will continue to write a for the radial polar coordinate.) The required average is

$$\frac{1}{\pi r^2} \int_0^{2\pi} \int_0^r \left(\frac{2}{\pi} \int_0^{\pi/2} \sqrt{r^2 - a^2 \sin^2 \alpha}\,d\alpha \right) a\,da\,d\theta$$

$$= \frac{4}{\pi r^2} \int_0^{\pi/2} \int_0^r a\sqrt{r^2 - a^2 \sin^2 \alpha}\,da\,d\alpha$$

$$= \frac{4}{\pi r^2} \int_0^{\pi/2} \left[-\frac{1}{3} \frac{(r^2 - a^2 \sin^2 \alpha)^{3/2}}{\sin^2 \alpha} \right]_{a=0}^{a=r} d\alpha$$

$$= \frac{4}{\pi r^2} \int_0^{\pi/2} \frac{-1}{3 \sin^2 \alpha} \left[(r^2 - r^2 \sin^2 \alpha)^{3/2} - (r^2)^{3/2} \right] d\alpha$$

$$= \frac{4}{\pi r^2} \int_0^{\pi/2} \frac{-1}{3 \sin^2 \alpha} (r^3 \cos^3 \alpha - r^3)\,d\alpha$$

$$= \frac{4r}{3\pi} \int_0^{\pi/2} \frac{1 - \cos^3 \alpha}{\sin^2 \alpha}\,d\alpha.$$

To continue with this improper integral, we first evaluate the indefinite integral.

$$\int \frac{1 - \cos^3 \alpha}{\sin^2 \alpha}\,d\alpha = \int \frac{1 - (1 - \sin^2 \alpha) \cos \alpha}{\sin^2 \alpha}\,d\alpha$$

$$= \int \left(\csc^2 \alpha - \frac{\cos \alpha}{\sin^2 \alpha} + \cos \alpha \right) d\alpha$$

$$= -\cot \alpha + \frac{1}{\sin \alpha} + \sin \alpha + C$$

$$= \frac{1 - \cos \alpha}{\sin \alpha} + \sin \alpha + C.$$

Therefore,

$$\int_0^{\pi/2} \frac{1-\cos^3 \alpha}{\sin^2 \alpha} \, d\alpha = \lim_{b \to 0+} \left[\frac{1-\cos \alpha}{\sin \alpha} + \sin \alpha \right]_b^{\pi/2}$$

$$= \lim_{b \to 0+} \left(1 + 1 - \frac{1-\cos b}{\sin b} - \sin b \right)$$

$$= 2,$$

and the average distance the insect crawls to the perimeter of the circle is

$$\frac{4r}{3\pi} \int_0^{\pi/2} \frac{1-\cos^3 \alpha}{\sin^2 \alpha} \, d\alpha = \frac{8}{3\pi} r.$$

Problem 100

Let $ABCD$ be a parallelogram in the plane. Describe and sketch the set of all points P in the plane for which there is an ellipse with the property that the points A, B, C, D, and P all lie on the ellipse.

Answer. The set of points P for which there is such an ellipse can be described by first extending all sides of the parallelogram indefinitely in all directions. Then P is in the set if and only if (0) P is one of the four points A, B, C, D, or (1) P is outside the parallelogram *and* (2) there is a pair of parallel extended sides of the parallelogram such that P is between those sides (see Figure 48).

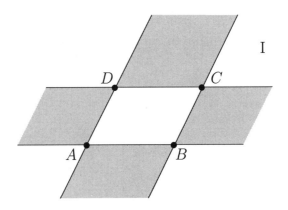

FIGURE 48

Solution 1. To show why this description is correct, choose coordinates in the plane such that $A = (0,0)$, $B = (1,0)$, and $D = (\lambda, \mu)$. Then $C = (\lambda + 1, \mu)$. We may assume that $\mu > 0$. For the four points $(0,0)$, $(1,0)$, (λ, μ), $(\lambda + 1, \mu)$ to lie on a conic section $ax^2 + bxy + cy^2 + dx + ey + f = 0$, we must have

$$f = 0,$$
$$a + d + f = 0,$$
$$a\lambda^2 + b\lambda\mu + c\mu^2 + d\lambda + e\mu + f = 0,$$
$$a(\lambda + 1)^2 + b(\lambda + 1)\mu + c\mu^2 + d(\lambda + 1) + e\mu + f = 0.$$

Since $f = 0$, the second, third, and fourth equations simplify to

$$a + d = 0, \tag{1}$$
$$a\lambda^2 + b\lambda\mu + c\mu^2 + d\lambda + e\mu = 0, \tag{2}$$
$$a(\lambda + 1)^2 + b(\lambda + 1)\mu + c\mu^2 + d(\lambda + 1) + e\mu = 0.$$

Subtracting these final two equations, one obtains $2a\lambda + a + b\mu + d = 0$, hence by (1),

$$2a\lambda + b\mu = 0. \tag{3}$$

Combining (1), (2), and (3), we find that $e = (a\lambda^2 - c\mu^2 + a\lambda)/\mu$. Thus, the constraints imply that any conic section through A, B, C, and D has the form

$$ax^2 - 2a\frac{\lambda}{\mu}xy + cy^2 - ax + \frac{a\lambda^2 - c\mu^2 + a\lambda}{\mu}y = 0 \tag{4}$$

for some real numbers a and c. Such a conic section is an ellipse if and only if its discriminant is negative. Therefore, a point $P = (x, y)$ is in the set if and only if there exist real numbers a and c such that (4) holds and $(-2a\lambda/\mu)^2 - 4ac < 0$; this inequality is equivalent to $a^2\lambda^2/\mu^2 < ac$. Clearly, $a \neq 0$, and dividing the equation of the conic section by a we get

$$x^2 - 2\frac{\lambda}{\mu}xy + \frac{c}{a}y^2 - x + \frac{\lambda^2 - (c/a)\mu^2 + \lambda}{\mu}y = 0.$$

If we now put $c_1 = c/a$, we see that P is in the set if and only if there exists a number c_1 such that

$$x^2 - 2\frac{\lambda}{\mu}xy + c_1y^2 - x + \frac{\lambda^2 - c_1\mu^2 + \lambda}{\mu}y = 0 \quad \text{and} \quad \frac{\lambda^2}{\mu^2} < c_1.$$

The equality can be rewritten as

$$c_1(y^2 - \mu y) = -x^2 + 2\frac{\lambda}{\mu}xy + x - \frac{\lambda^2 + \lambda}{\mu}y. \tag{5}$$

We now distinguish three cases.

Case 1. $y^2 - \mu y = 0$. Then $y = 0$ or $y = \mu$. If $y = 0$, the equality yields $x = 0$ or $x = 1$, and we find that $P = A$ or $P = B$.

Similarly, for $y = \mu$ we find that $P = C$ or $P = D$.

Case 2. $y^2 - \mu y > 0$. In this case $\lambda^2/\mu^2 < c_1$ is equivalent to

$$(\lambda^2/\mu^2)(y^2 - \mu y) < c_1(y^2 - \mu y),$$

and this together with (5) shows that P is in the set if and only if the following equivalent inequalities hold:

$$\frac{\lambda^2}{\mu^2}(y^2 - \mu y) < -x^2 + 2\frac{\lambda}{\mu}xy + x - \frac{\lambda^2 + \lambda}{\mu}y,$$

$$\lambda^2 y^2 < -\mu^2 x^2 + 2\lambda\mu xy + \mu^2 x - \lambda\mu y,$$

$$(\lambda y - \mu x)(\lambda y - \mu x + \mu) < 0,$$

$$\lambda y - \mu x < 0 \quad \text{and} \quad \lambda y - \mu x + \mu > 0,$$

$$-\mu < \lambda y - \mu x < 0.$$

(Note that since $\mu > 0$, it would be impossible to have $\lambda y - \mu x > 0$ and $\lambda y - \mu x + \mu < 0$.)

Now it is easy to check that $\lambda y - \mu x = 0$ is the equation of line AD, while $\lambda y - \mu x = -\mu$ is the equation of BC. So the condition $-\mu < \lambda y - \mu x < 0$ implies that P is between (the extensions of) AD and BC. On the other hand, $y^2 - \mu y > 0$ is equivalent to ($y < 0$ or $y > \mu$), which says that P is not between AB and CD.

Case 3. $y^2 - \mu y < 0$. The same computation, but with the inequalities reversed, shows that P is not between the lines AD and BC, while P is between (the extensions of) AB and CD. Combining the results of the three cases, we get the answer given above.

Solution 2. Since the set consisting of an ellipse and its interior is convex, it is impossible for a point on an ellipse to lie inside the triangle determined by three other points on that ellipse. We see from this that only the points P described above can be on the same ellipse as A, B, C, and D. For instance, if P is in the region labeled "I" in Figure 48, then C is inside triangle BDP, so P cannot be on any ellipse on which B, C, and D lie.

To see the converse, choose a coordinate system for the plane such that $A = (0,0)$, $B = (1,0)$, $C = (\lambda + 1, \mu)$, and $D = (\lambda, \mu)$, where $\mu \neq 0$. The linear transformation T taking (x, y) to $(x - (\lambda/\mu)y, (1/\mu)y)$ takes A, B, C, and D to $(0,0)$, $(1,0)$, $(1,1)$, and $(0,1)$, respectively. Since both T and T^{-1} map straight lines to straight lines and ellipses to ellipses, we see that it suffices to consider the case $A = (0,0)$, $B = (1,0)$, $C = (1,1)$, and $D = (0,1)$. In this case the points $P = (r, s)$, other than A, B, C, D, described above lie in two horizontal and two vertical "half-strips" given by $(r - r^2)/(s^2 - s) > 0$. For such a point P, the curve

$$(x^2 - x) + \frac{r - r^2}{s^2 - s}(y^2 - y) = 0$$

is an ellipse (in fact, the unique ellipse) through A, B, C, D, and P.

Problem 101

Let x_0 be a rational number, and let $(x_n)_{n \geq 0}$ be the sequence defined recursively by

$$x_{n+1} = \left| \frac{2x_n^3}{3x_n^2 - 4} \right|.$$

Prove that this sequence converges, and find its limit as a function of x_0.

Solution.

$$\lim_{n \to \infty} x_n = \begin{cases} 0 & \text{if } |x_0| < 2/\sqrt{5}, \\ 2 & \text{if } |x_0| > 2/\sqrt{5}. \end{cases}$$

Note that all x_n are actually rational numbers, since when x_n is rational, $3x_n^2 - 4$ will not be zero and so x_{n+1} will again be rational. Also note that after the initial term, the sequence starting with x_0 is the same as the sequence starting with $-x_0$, so we can assume $x_0 \geq 0$.

Let $f(x) = 2x^3/(3x^2 - 4)$, so that $x_{n+1} = |f(x_n)|$. We can use standard techniques to sketch the graph of f (shown in Figure 49 for $x \geq 0$ only). From this graph it seems reasonable to suspect that $2 < f(x) < x$ for $x > 2$, and it is straightforward to show this. Thus, if $x_n > 2$ for some n, the sequence will thereafter decrease, but stay above 2. Since any bounded decreasing sequence has a limit, $L = \lim_{n \to \infty} x_n$ will then exist. Taking limits on both sides of $x_{n+1} = f(x_n)$ and using the continuity of f for $x \geq 2$, we then have $L = f(L)$, so $L(3L^2 - 4) = 2L^3$ and $L(L^2 - 4) = 0$. Of the three roots, $L = 0$, $L = \pm 2$, only $L = 2$ is possible in this case, since $x_n > 2$ for large enough n. In particular,

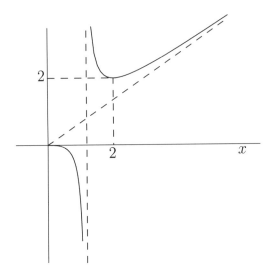

FIGURE 49

we see that $L = 2$ if $x_0 > 2$. We can also see that $L = 2$ if $x_0 = 2$ (in this case, $x_n = 2$ for all n) and even that $L = 2$ if $2/\sqrt{3} < x_0 < 2$, for in that case $x_1 = f(x_0) > 2$.

Now consider what happens if $0 \le x_0 < 2/\sqrt{3}$. If $x_0 = 0$, then $x_n = 0$ for all n, so $L = 0$. From the graph, we expect that for x close to 0,

$$|f(x)| = \frac{2x^3}{4 - 3x^2}$$

will be even closer to 0, while for x close enough to $2/\sqrt{3}$, $|f(x)|$ becomes large and thus $|f(x)| > x$.

A simple calculation shows that $2x^3/(4 - 3x^2) = x$ for $x = 0$ and for $x = \pm 2/\sqrt{5}$. If $0 < x_n < 2/\sqrt{5}$ at any point in the sequence, then it is easily shown that $0 < x_{n+1} = |f(x_n)| < x_n$, and so forth, so the sequence will decrease (and be bounded) from that point on. Thus, L will exist; as before, we have $L = |f(L)| = 2L^3/(4 - 3L^2)$, which implies $L = 0$ or $L = \pm 2/\sqrt{5}$. However, once $0 < x_n < 2/\sqrt{5}$ and (x_n) is decreasing, we must have $L = 0$.

Since x_0 is rational, $x_0 = 2/\sqrt{5}$ is impossible, so we are left with the case $2/\sqrt{5} < x_0 < 2/\sqrt{3}$. In this case, $x_1 = |f(x_0)| > x_0$, so the sequence increases at first. Suppose that $x_n < 2/\sqrt{3}$ for all n. In this case, the sequence is bounded and increasing, so L exists. But from $L = |f(L)|$ and $0 < L \le 2/\sqrt{3}$ we again get $L = 0$ or $L = \pm 2/\sqrt{5}$, a contradiction since $x_0 > 2/\sqrt{5}$ and (x_n) is increasing. It follows that there is an n with $x_n > 2/\sqrt{3}$. But then $x_{n+1} \geq 2$, so $L = 2$, and we are done.

Comment. The stipulation that x_0 be rational is stronger than necessary. However, some restriction is needed to ensure that x_{n+1} is defined for all n.

Problem 102

Let f be a continuous function on $[0, 1]$, which is bounded below by 1, but is not identically 1. Let R be the region in the plane given by $0 \leq x \leq 1, 1 \leq y \leq f(x)$. Let

$$R_1 = \{(x, y) \in R \big| y \leq \bar{y}\} \quad \text{and} \quad R_2 = \{(x, y) \in R \big| y \geq \bar{y}\},$$

where $\bar{y}$ is the y-coordinate of the centroid of R. Can the volume obtained by rotating R_1 about the x-axis equal that obtained by rotating R_2 about the x-axis?

Answer. Yes, the volumes can be equal. For example, this happens if

$$f(x) = \frac{22}{13} - \frac{9}{13}x.$$

Solution 1. To find an example, we let R be the triangle with vertices $(0, 1)$, $(1, 1)$, and $(0, b)$, where the parameter $b > 1$ will be chosen later. Then

$$f(x) = b - (b - 1)x.$$

There are several expressions for $\bar{y}$ in terms of integrals. For instance,

$$\bar{y} = \frac{\displaystyle\int_1^b y \frac{b - y}{b - 1} \, dy}{\displaystyle\int_1^b \frac{b - y}{b - 1} \, dy} = \frac{b + 2}{3}.$$

The horizontal line $y = (b + 2)/3$ intersects $y = f(x)$ at $(2/3, (b + 2)/3)$. The volume obtained by rotating R_2 about the x-axis is then

$$\int_0^{2/3} \left(\pi \left(f(x)\right)^2 - \pi \left(\frac{b + 2}{3}\right)^2 \right) dx = \frac{20b^2 - 4b - 16}{81} \pi.$$

To find the volume obtained by rotating R_1 about the x–axis, we compute the volume obtained by rotating R, and subtract the volume obtained by rotating

R_2. The calculation yields

$$\frac{7b^2 + 31b - 38}{81} \pi.$$

The two volumes are equal when $b = 22/13$ and when $b = 1$, but the solution $b = 1$ is degenerate.

Solution 2. As in the first solution, we let R be the triangle with vertices $(0, 1)$, $(1, 1)$, and $(0, b)$, but this time we avoid computing integrals by using geometry. Recall that by Pappus' theorem, the volume swept out by rotating a region about an exterior axis is the product of the area of the region and the distance traveled by its centroid. Therefore, if $\bar{y}_1$ and $\bar{y}_2$ are the y-coordinates of the centroids of R_1 and R_2 respectively, the two volumes will be equal if and only if

$$(\text{Area of } R_1) \cdot 2\pi\bar{y}_1 = (\text{Area of } R_2) \cdot 2\pi\bar{y}_2.$$

Now R is a triangle, so its centroid, being the intersection of its medians, is two-thirds of the way down from $(0, b)$ to the line $y = 1$; that is, $\bar{y} = (b + 2)/3$. Similarly, for the triangle R_2 we have

$$\bar{y}_2 = \frac{b + 2\bar{y}}{3} = \frac{5b + 4}{9}.$$

Also, $\bar{y}$ is the weighted average of $\bar{y}_1$ and $\bar{y}_2$, where the respective weights are the areas of R_1 and R_2. By similar triangles, the area of R_2 is $4/9$ times the area of R, so

$$\bar{y} = \frac{5}{9}\bar{y}_1 + \frac{4}{9}\bar{y}_2 ;$$

this yields $\bar{y}_1 = (7b + 38)/45$. The condition for the volumes to be equal now becomes

$$\frac{5}{9} \cdot \frac{7b + 38}{45} = \frac{4}{9} \cdot \frac{5b + 4}{9},$$

and we again find $b = 22/13$.

Comment. These solutions illustrate a useful method of constructing examples and counterexamples: take a parametrized family for which computations are feasible, and then choose the parameter(s) so as to get the desired properties.

Problem 103

Let $n \geq 3$ be a positive integer. Begin with a circle with n marks about it. Starting at a given point on the circle, move clockwise, skipping over the next two marks and placing a new mark; the circle now has $n+1$ marks. Repeat the procedure beginning at the new mark. Must a mark eventually appear between each pair of the original marks?

Solution.   Yes, a mark must eventually appear between any pair of original marks.

To show why, suppose there is a pair m_1, m_2 of consecutive original marks such that no mark will ever appear between them. This implies that as you move around the circle, you always put a mark just before m_1, then skip over both m_1 and m_2 and put the next mark right after m_2, as shown in Figure 50.

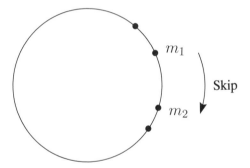

FIGURE 50

Now suppose that for some particular time around the circle, there are x marks in all as you skip m_1 and m_2. On your next circuit (counting from that moment), you must again arrive just before m_1 in order to skip over m_1 and m_2, so x must be even and you will be adding $x/2$ marks during this next circuit, for a total of $3x/2$. However, not all integers in the sequence $x, (3/2)x, (3/2)^2 x, \ldots$ can be even. Thus on some circuit there will be an odd number of marks as you skip m_1 and m_2, and on the next circuit a mark will appear between m_1 and m_2, a contradiction.

Comments.   One can see explicitly when a mark will appear between any two original marks, as follows. Label the original marks consecutively from 1 to n, and let I_a be the interval between the mark labeled a and the next mark; suppose that the starting position is in interval I_n. Label the new marks beginning with $n+1$. Then if $n = 2^j k$, where k is odd, mark $(3^{j+1}k+1)/2$ will be the first one to fall in interval I_1, and one must go around the circle $j+1$ times before this happens.

More generally, for $1 \le a \le n$, the number of the first mark to fall in interval I_a is $P(n, a)$, where

$$P(n, a) = \begin{cases} n + a/2, & \text{if } a \text{ is even}; \\ (3^{j+1}k + 1)/2, & \text{if } a \text{ is odd}, n + (a-1)/2 = 2^j k, k \text{ odd}. \end{cases}$$

Replacing "two marks" by "m marks" (and "$n \ge 3$" by "$n \ge m+1$") in the statement of the problem yields an interesting generalization, which we have not yet been able to solve.

Problem 104

Let

$$c = \sum_{n=1}^{\infty} \frac{1}{n(2^n - 1)} = 1 + \frac{1}{6} + \frac{1}{21} + \frac{1}{60} + \cdots.$$

Show that

$$e^c = \frac{2}{1} \cdot \frac{4}{3} \cdot \frac{8}{7} \cdot \frac{16}{15} \cdots.$$

Solution. Let P_k be the partial product

$$\frac{2}{1} \cdot \frac{4}{3} \cdot \frac{8}{7} \cdots \frac{2^k}{2^k - 1}.$$

We want to show that

$$\lim_{k \to \infty} P_k = e^c,$$

where

$$c = \sum_{n=1}^{\infty} \frac{1}{n(2^n - 1)}.$$

By continuity of the exponential and logarithm functions, this is equivalent to showing that $\lim_{k\to\infty} \ln(P_k) = c$. Now

$$\ln(P_k) = \ln\left(\frac{2}{1} \cdot \frac{4}{3} \cdot \frac{8}{7} \cdots \frac{2^k}{2^k-1}\right)$$

$$= \ln\left(\frac{2}{1}\right) + \ln\left(\frac{4}{3}\right) + \ln\left(\frac{8}{7}\right) + \cdots + \ln\left(\frac{2^k}{2^k-1}\right)$$

$$= -\ln\left(\frac{1}{2}\right) - \ln\left(\frac{3}{4}\right) - \ln\left(\frac{7}{8}\right) - \cdots - \ln\left(\frac{2^k-1}{2^k}\right)$$

$$= -\ln\left(1-\frac{1}{2}\right) - \ln\left(1-\frac{1}{4}\right) - \ln\left(1-\frac{1}{8}\right) - \cdots - \ln\left(1-\frac{1}{2^k}\right)$$

$$= \sum_{i=1}^{k} -\ln\left(1-\frac{1}{2^i}\right).$$

Using the Maclaurin expansion (Taylor expansion about 0)

$$-\ln(1-x) = x + \frac{x^2}{2} + \frac{x^3}{3} + \cdots = \sum_{n=1}^{\infty} \frac{x^n}{n} \qquad (|x| < 1),$$

we get

$$\ln(P_k) = \sum_{i=1}^{k}\sum_{n=1}^{\infty} \frac{(1/2^i)^n}{n} = \sum_{n=1}^{\infty}\sum_{i=1}^{k} \frac{1}{n2^{in}} = \sum_{n=1}^{\infty}\frac{1}{n}\sum_{i=1}^{k}\frac{1}{(2^n)^i}.$$

Now

$$\sum_{i=1}^{k} \frac{1}{(2^n)^i} = \frac{1}{2^n} + \frac{1}{(2^n)^2} + \cdots + \frac{1}{(2^n)^k}$$

is a finite geometric series with sum

$$\frac{1}{2^n} \cdot \frac{1-(1/2^n)^k}{1-1/2^n} = \frac{1-(1/2^n)^k}{2^n-1},$$

so

$$\ln(P_k) = \sum_{n=1}^{\infty} \frac{1-(1/2^n)^k}{n(2^n-1)} = \sum_{n=1}^{\infty} \frac{1}{n(2^n-1)} - \sum_{n=1}^{\infty} \frac{1}{n(2^n-1)2^{nk}}$$

since the latter two series converge. The estimate

$$\sum_{n=1}^{\infty} \frac{1}{n(2^n-1)2^{nk}} < \frac{1}{2^k}\sum_{n=1}^{\infty} \frac{1}{n(2^n-1)}$$

allows us to conclude that

$$\lim_{k \to \infty} \ln\left(P_k\right) = \sum_{n=1}^{\infty} \frac{1}{n(2^n - 1)} = c,$$

as claimed.

Problem 105

Let $q(x) = x^2 + ax + b$ be a quadratic polynomial with real roots. Must all roots of $p(x) = x^3 + ax^2 + (b-3)x - a$ be real?

Answer. Yes, the roots of $p(x) = x^3 + ax^2 + (b-3)x - a$ will all be real if the roots of $q(x) = x^2 + ax + b$ are real.

Solution 1. Note that if we replace x by $-x$, we get the same problem with a replaced by $-a$; thus, we may assume that $a \geq 0$. Let $r_1 \leq r_2$ be the roots of $q(x)$. We then have $r_1 + r_2 = -a$, hence $r_1 \leq 0$. If we note that $p(x) = xq(x) - (3x + a)$, we find $p(r_1) = -3r_1 - a = -2r_1 + r_2 \geq 0$. If $a > 0$, then from $p(0) = -a < 0$ and the Intermediate Value Theorem, $p(x)$ has a negative and a positive root, hence a third real root as well. If $a = 0$, then in order for $x^2 + b$ to have real roots, we must have $b \leq 0$. But then all roots of the polynomial $p(x) = x^3 + (b-3)x$ will be real.

Solution 2. It is easy to check that

$$p(x) = \mathrm{Re}\big((x+i)q(x+i)\big) = \big((x+i)q(x+i) + (x-i)q(x-i)\big)/2.$$

If z is a (complex) root of $p(x)$, then, in particular,

$$\big|(z+i)\,q(z+i)\big| = \big|(z-i)\,q(z-i)\big|,$$

or, if we denote the (real) roots of $q(x)$ by r_1 and r_2,

$$\big|(z+i)\,(z+i-r_1)\,(z+i-r_2)\big| = \big|(z-i)\,(z-i-r_1)\,(z-i-r_2)\big|.$$

If z has positive imaginary part, then each of the three factors on the left-hand side of the equation will be larger in absolute value than the corresponding factor on the right-hand side, a contradiction. Similarly, z cannot have negative imaginary part; hence, z must be real.

Solution 3. Note that

$$p(x) = x(x^2 + ax + b) - (2x + a) - x$$
$$= xq(x) - q'(x) - x;$$

we will use this connection between $p(x)$ and $q(x)$ to show that all roots of $p(x)$ are real. Let r_1 and r_2, with $r_1 \le r_2$, be the roots of $q(x)$. Since the graph of $q(x)$ opens upward, we have $q'(r_1) \le 0$ and $q'(r_2) \ge 0$, with equality only when $r_1 = r_2$. We now distinguish three cases.

Case 1. $r_1 \le 0$, $r_2 \ge 0$. In this case, $p(r_1) = -q'(r_1) - r_1 \ge 0$ and $p(r_2) \le 0$. Thus, by the Intermediate Value Theorem, $p(x)$ has a root in the interval $[r_1, r_2]$. Since $\lim_{x \to -\infty} p(x) = -\infty$ and $\lim_{x \to \infty} p(x) = \infty$, $p(x)$ also has roots in each of the intervals $(-\infty, r_1]$ and $[r_2, \infty)$, so we are done with *Case 1* unless $p(r_1) = 0$ or $p(r_2) = 0$. If $p(r_1) = 0$, we must have $q'(r_1) = r_1 = 0$, so $r_1 = r_2 = 0$, hence $q(x) = x^2$ and $p(x) = x^3 - 3x$, with all roots real. The case $p(r_2) = 0$ is similar.

Case 2. $r_1 > 0$, $r_2 > 0$. Since $q(x) = (x - r_1)(x - r_2)$, we have $a = -r_1 - r_2 < 0$ and so $p(0) = -a > 0$. On the other hand, $p(r_2) = -q'(r_2) - r_2 < 0$, so $p(x)$ has roots in each of the intervals $(-\infty, 0)$, $(0, r_2)$, and (r_2, ∞).

Case 3. $r_1 < 0$, $r_2 < 0$. This time, $a = -r_1 - r_2 > 0$, so $p(0) < 0$, while $p(r_1) = -q'(r_1) - r_1 > 0$. Thus $p(x)$ has roots in each of the intervals $(-\infty, r_1)$, $(r_1, 0)$, and $(0, \infty)$, and we are done.

Problem 106

Let $p(x) = x^3 + a_1 x^2 + a_2 x + a_3$ have rational coefficients and have roots r_1, r_2, r_3. If $r_1 - r_2$ is rational, must r_1, r_2, and r_3 be rational?

Answer. Yes, if $r_1 - r_2$ is rational, then r_1, r_2, and r_3 must all be rational.

Solution 1. Let $q = r_1 - r_2$ be rational. Note that if either r_1 or r_2 is rational, then so is the other, and since $r_1 + r_2 + r_3 = -a_1$ is rational, so is r_3. On the other hand, if r_3 is rational, then so is $r_1 + r_2 = -a_1 - r_3$, hence $r_1 = ((r_1 + r_2) + (r_1 - r_2))/2$ and r_2 are rational also. Thus it is enough to show that $p(x)$ has at least *one* rational root. We do this by showing that $p(x)$ is reducible over the rationals; for then, since it is a cubic polynomial, it must have a linear factor, and therefore a rational root.

Using the rational numbers $q = r_1 - r_2$ and $-a_1 = r_1 + r_2 + r_3$, we can express the roots r_2 and r_3 as $r_2 = r_1 - q$, $r_3 = -2r_1 + q - a_1$. We then have

$$a_2 = r_1 r_2 + r_1 r_3 + r_2 r_3$$
$$= r_1(r_1 - q) + r_1(-2r_1 + q - a_1) + (r_1 - q)(-2r_1 + q - a_1)$$
$$= -3r_1^2 + q_1 r_1 + q_2,$$

where q_1 and q_2 are rational. Therefore, r_1 is a root of the rational quadratic polynomial $p_1(x) = 3x^2 - q_1 x + (a_2 - q_2)$. Since $p(x)$ and $p_1(x)$ have the root r_1 in common, their greatest common divisor $d(x)$, which has rational coefficients (and can be found using the Euclidean algorithm), also has r_1 as a root. Hence $d(x)$, which has degree 1 or 2, is a divisor of $p(x)$, so $p(x)$ is reducible over the rationals, and we are done.

Solution 2. (Eugene Luks, University of Oregon)   As in Solution 1, it is enough to show that $p(x)$ is reducible over the rationals. Consider the polynomial $p_1(x) = p(x + r_1 - r_2)$. Since $r_1 - r_2$ is rational, $p_1(x)$ has rational coefficients. Also, $p_1(r_2) = p(r_1) = 0$, so $p(x)$ and $p_1(x)$ have r_2 as a common root. Thus the greatest common divisor $d(x)$ of $p(x)$ and $p_1(x)$ is a (monic) rational polynomial of degree at least 1, which shows that $p(x)$ is reducible unless $p(x) = d(x) = p_1(x)$. If $p(x) = p_1(x)$, then $r_1 = r_2$ and $p(x)$ has a multiple root r_1. But then r_1 is also a root of $p'(x)$, so the greatest common divisor of $p(x)$ and $p'(x)$, which is rational, has degree 1 or 2, so once again $p(x)$ is reducible over the rationals.

Solution 3.   As in Solution 1, it is enough to show that $p(x)$ is reducible over the rationals. Suppose not. Then by Galois theory, the splitting field $\mathbf{Q}(r_1, r_2, r_3)$ of $p(x)$ has an automorphism σ which takes r_1 to r_2, r_2 to r_3, and r_3 to r_1. Now $r_1 - r_2$ is rational and hence fixed by σ, so $r_1 - r_2 = \sigma(r_1) - \sigma(r_2) = r_2 - r_3$, that is, $r_3 = 2r_2 - r_1$. On the other hand, $r_1 + r_2 + r_3 = -a_1$ is rational, so $r_1 + r_2 + (2r_2 - r_1) = 3r_2$ is rational, r_2 is rational, and we are done.

Comment.   One natural generalization to polynomials of degree 4 does not hold. That is, there exist fourth-degree rational polynomials with irrational roots r_1, r_2, r_3, r_4 for which two differences of pairs of roots are rational; one example is $p(x) = x^4 + 4$, with roots $1 + i$, $-1 + i$, $1 - i$, $-1 - i$ and $r_1 - r_2 = r_3 - r_4 = 2$.

Problem 107

Let $f(x) = x^3 - 3x + 3$. Prove that for any positive integer P, there is a "seed" value x_0 such that the sequence $x_0, x_1, x_2, \ldots$ obtained from Newton's method, given by

$$x_{n+1} = x_n - \frac{f(x_n)}{f'(x_n)},$$

has period P.

Idea. If the "seed" is planted at a point just to the right of the critical point $x = -1$, the following term in the sequence will be a large positive number and the next few terms will decrease (see Figure 51). If the seed is planted exactly right, the sequence might return to the seed after P steps.

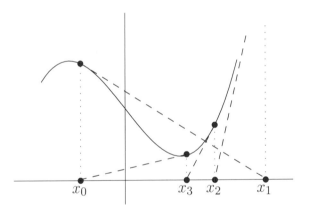

FIGURE 51

Solution. We first consider the special case $P = 1$. For the sequence to have period 1, we need $f(x_0) = 0$. Since any cubic polynomial with real coefficients has a real root, we can indeed choose x_0 such that $f(x_0) = 0$; we then have $x_1 = x_0$, the sequence has period 1, and we are done.

From now on we assume $P > 1$. Let

$$F(x) = x - \frac{f(x)}{f'(x)} = \frac{2x^3 - 3}{3(x^2 - 1)},$$

so that the sequence is x_0, $x_1 = F(x_0)$, $x_2 = F(F(x_0))$, ... We have

$$F'(x) = \frac{f''(x)f(x)}{(f'(x))^2} = \frac{6x}{(3(x^2-1))^2}(x^3 - 3x + 3).$$

Since $f(x)$ increases on the interval $[1, \infty)$, it is at least 1 there, so we see that $F'(x) > 0$ on $(1, \infty)$. Thus, F is increasing on $(1, \infty)$. Note that

$$\lim_{x \to 1+} F(x) = -\infty \qquad \text{and} \qquad \lim_{x \to \infty} F(x) = \infty.$$

Therefore, if F is restricted to $(1, \infty)$, it has a continuous inverse G whose domain is $(-\infty, \infty)$. Thus, for any real number x, $G(x)$ is the unique number in $(1, \infty)$ for which $F(G(x)) = x$. (Note that to go from x to $G(x)$ is "backtracking" in Newton's method.) Let G^k denote the composition of k copies of G; that is,

$$G^k(x) = \underbrace{G \circ G \circ \cdots \circ G}_{k \text{ times}}(x) = \underbrace{G(G(\cdots(G(x))\cdots))}_{k \text{ times}}.$$

Define $H(x)$ on $(-1, 0]$ by $H(x) = F(x) - G^{P-1}(x)$. Since F and G are continuous on $(-1, 0]$, so is H, and since $G^{P-1}(-1)$ is finite,

$$\lim_{x \to -1+} H(x) = \infty.$$

Meanwhile, $F(0) = 1$ and all values of G are greater than 1, so $H(0) < 0$. Therefore, by the Intermediate Value Theorem there exists an $x_0 \in (-1, 0)$ for which $H(x_0) = 0$. We claim that the sequence starting from this seed value x_0 has period P. First, since $H(x_0) = 0$, we have $F(x_0) = G^{P-1}(x_0)$ and thus

$$x_P = F^P(x_0) = F^{P-1}(G^{P-1}(x_0)) = x_0.$$

To show that the sequence cannot have a period less than P, note that $F(x) < x$ for all x in $(1, \infty)$, so $G(x) > x$ for all x. Therefore,

$$x_1 = G^{P-1}(x_0) > x_2 = G^{P-2}(x_0) > \cdots > x_{P-1} = G(x_0) > x_0,$$

and x_P is the first term of the sequence to return to x_0.

Problem 108

Show that

$$\sum_{k=0}^{n} \frac{(-1)^k}{2n + 2k + 1}\binom{n}{k} = \frac{(2^n(2n)!)^2}{(4n+1)!}.$$

Solution. Starting with the left-hand side, we have

$$\sum_{k=0}^{n} \binom{n}{k} \frac{(-1)^k}{2n+2k+1} = \sum_{k=0}^{n} \binom{n}{k}(-1)^k \int_0^1 u^{2n+2k}\,du$$

$$= \int_0^1 u^{2n} \sum_{k=0}^{n} \binom{n}{k}(-u^2)^k\,du$$

$$= \int_0^1 u^{2n}(1-u^2)^n\,du.$$

To evaluate this integral (which is available from tables, see the comments below), we generalize it by defining

$$F(k,m) = \int_0^1 u^k(1-u^2)^m\,du;$$

note that we are looking for $F(2n,n)$. Now $F(k,0) = 1/(k+1)$, and integration by parts yields

$$F(k,m) = \frac{2m}{k+1} F(k+2, m-1)$$

for $m > 0$. Applying this recursive equation repeatedly, we find

$$F(2n,n) = \frac{2^n\, n!}{(2n+1)(2n+3)\cdots(4n-1)}\, F(4n,0)$$

$$= \frac{2^n\, n!}{(2n+1)(2n+3)\cdots(4n-1)(4n+1)}$$

$$= \frac{(2n)!}{(2n)!} \cdot \frac{2^n\, n!}{(2n+1)(2n+3)\cdots(4n-1)(4n+1)}$$

$$\cdot \frac{2^n(n+1)(n+2)\cdots(2n)}{(2n+2)(2n+4)\cdots(4n)}$$

$$= \frac{\left(2^n\,(2n)!\right)^2}{(4n+1)!},$$

as was to be shown.

Comments. There are several other approaches to finding the integral

$$I = \int_0^1 u^{2n}(1-u^2)^n\,du\,.$$

One is to convert I into a "standard" integral (found in tables) by the substitution $u^2 = x$. This yields

$$I = \tfrac{1}{2} \int_0^1 x^{n-\frac{1}{2}} (1-x)^n \, dx = \tfrac{1}{2} \, \mathrm{B}(n+\tfrac{1}{2}, n+1),$$

where B is the *beta function* defined by

$$\mathrm{B}(s,t) = \int_0^1 x^{s-1}(1-x)^{t-1} \, dx, \qquad s, t > 0.$$

It is known that

$$\mathrm{B}(s,t) = \frac{\Gamma(s)\Gamma(t)}{\Gamma(s+t)}.$$

Here Γ is the *gamma function*:

$$\Gamma(s) = \int_0^\infty x^{s-1} e^{-x} dx, \qquad s > 0.$$

This last function is known to have the specific values

$$\Gamma(n+\tfrac{1}{2}) = \frac{(2n)!}{4^n \, n!} \sqrt{\pi}, \qquad n = 1, 2, 3, \ldots,$$

$$\Gamma(n) = (n-1)!, \qquad n = 1, 2, 3, \ldots,$$

and so we get

$$I = \tfrac{1}{2} \, \mathrm{B}(n+\tfrac{1}{2}, n+1)$$

$$= \frac{1}{2} \frac{\Gamma(n+\tfrac{1}{2})\Gamma(n+1)}{\Gamma(2n+\tfrac{3}{2})}$$

$$= \frac{1}{2} \frac{\dfrac{(2n)!}{4^n \, n!}\sqrt{\pi}\, n!}{\dfrac{(4n+2)!}{4^{2n+1}(2n+1)!}\sqrt{\pi}}$$

$$= \frac{\left(2^n \, (2n)!\right)^2}{(4n+1)!}.$$

Another approach, which essentially derives the expression for the beta function in terms of gamma functions, starts with the substitution $u = \sin\theta$, which yields

$$I = \int_0^{\pi/2} \sin^{2n}\theta \, \cos^{2n+1}\theta \, d\theta.$$

By considering the double integral

$$\iint_R (r \sin \theta)^{2n} (r \cos \theta)^{2n+1} e^{-r^2} r \, dr \, d\theta,$$

where R is the first quadrant, and converting it to rectangular coordinates, one can then find that

$$I \cdot \int_0^\infty r^{4n+2} e^{-r^2} \, dr = \iint_R (r \sin \theta)^{2n} (r \cos \theta)^{2n+1} e^{-r^2} r \, dr \, d\theta$$

$$= \int_0^\infty y^{2n} e^{-y^2} \, dy \cdot \int_0^\infty x^{2n+1} e^{-x^2} \, dx \, .$$

Changing all integration variables to t, we have

$$I = \frac{\displaystyle\int_0^\infty t^{2n} e^{-t^2} \, dt \cdot \int_0^\infty t^{2n+1} e^{-t^2} \, dt}{\displaystyle\int_0^\infty t^{4n+2} e^{-t^2} \, dt}$$

$$= \frac{I_{2n} \, I_{2n+1}}{I_{4n+2}} \, ,$$

where $I_k = \displaystyle\int_0^\infty t^k e^{-t^2} \, dt$.

Now a straightforward substitution yields $I_1 = \frac{1}{2}$, and integration by parts shows that $I_k = ((k-1)/2) \cdot I_{k-2}$ for $k \geq 2$. Therefore,

$$I = \frac{I_{2n} \, I_{2n+1}}{I_{4n+2}} = \frac{I_{2n} \cdot n \cdot (n-1) \cdots 1 \cdot I_1}{\dfrac{4n+1}{2} \cdot \dfrac{4n-1}{2} \cdots \dfrac{2n+1}{2} \cdot I_{2n}}$$

$$= \frac{n! \, 2^n}{(4n+1)(4n-1) \cdots (2n+1)} \, ,$$

as in the solution above.

Problem 109

Suppose a and b are distinct real numbers such that

$$a - b, \, a^2 - b^2, \, \ldots, a^k - b^k, \, \ldots$$

are all integers.
a. Must a and b be rational?
b. Must a and b be integers?

Solution. Yes, a and b must be rational, and in fact, a and b must be integers.

a. Since a and b are distinct, we know that $a + b = (a^2 - b^2)/(a - b)$ is a rational number. Therefore, $a = ((a + b) + (a - b))/2$ is rational and hence b is also rational.

b. Since $a - b$ is an integer, a and b must have the same denominator, say n, when expressed in lowest terms. If we put $a = c/n$ and $b = d/n$, we see that n^k divides $c^k - d^k$ for $k = 1, 2, \ldots$.

If $n = 1$, we are done; otherwise, let p be a prime factor of n. Since n divides $c - d$, we have $c \equiv d \pmod{p}$. Now note that

$$c^k - d^k = (c - d)(c^{k-1} + c^{k-2}d + \cdots + cd^{k-2} + d^{k-1}),$$

and since $c \equiv d \pmod{p}$, the second factor is

$$c^{k-1} + c^{k-2}d + \cdots + cd^{k-2} + d^{k-1} \equiv kd^{k-1} \pmod{p}.$$

We have $d \not\equiv 0 \pmod{p}$ since $b = d/n$ is in lowest terms, so provided k is not divisible by p, we have $kd^{k-1} \not\equiv 0 \pmod{p}$. Thus p does not divide the second factor above of $c^k - d^k$, and since p^k divides $c^k - d^k$, p^k must divide $c - d$. However, we can choose arbitrarily large k which are not divisible by p, and so we have a contradiction.

Problem 110

The mayor of Wohascum Center has ten pairs of dress socks, ranging through ten shades of color from medium gray (1) to black (10). A pair of socks is unacceptable for wearing if the colors of the two socks differ by more than one shade. What is the probability that if the socks get paired at random, they will be paired in such a way that all ten pairs are acceptable?

Idea. Acceptable pairings can be built up by starting with an extreme color.

Solution. The probability is

$$\frac{683}{19 \cdot 17 \cdot 15 \cdots 3} \approx 10^{-6}.$$

To show this, we first show that there are 683 acceptable ways to pair the socks. Suppose we had n pairs of socks in gradually darkening shades, with the same condition for a pair to be acceptable. Let $f(n)$ be the number of pairings into acceptable pairs. For $n \leq 2$, any pairing is acceptable, and we have $f(1) = 1$, $f(2) = 3$. For $n > 2$, consider one sock from pair n (the darkest pair). It must be paired with its match or with one of the socks in pair $n - 1$. If it is paired with its match, there are $f(n - 1)$ acceptable ways to pair the remaining $n - 1$ pairs. If it is paired with either of the two socks from pair $n - 1$, its match must be paired with the other sock from that pair, which leaves $n - 2$ matching pairs and $f(n - 2)$ acceptable ways to pair them. Since there are two ways to do this, we conclude that, for $n > 2$,

$$f(n) = f(n - 1) + 2 f(n - 2).$$

The characteristic equation of this recurrence relation is $r^2 - r - 2 = 0$, which has solutions $r = 2$ and $r = -1$. Therefore, we have

$$f(n) = a \, 2^n + b \, (-1)^n$$

for some constants a and b. From $f(1) = 1$ and $f(2) = 3$, we find $a = 2/3$, $b = 1/3$, so

$$f(n) = \frac{2^{n+1} + (-1)^n}{3}.$$

In particular, the number of acceptable ways to pair 10 pairs of socks is $f(10) = 683$.

To compute the probability that a random pairing is acceptable, we divide $f(10)$ by the total number of pairings. To find the total number of pairings, note that a given sock can be paired in 19 ways. One given sock of the remaining 18 can be paired in 17 ways, and so forth. Thus the total number of pairings is $19 \cdot 17 \cdot 15 \cdots 3 \cdot 1$, and we get the probability given above.

Comment. To solve the recursion $f(n) = f(n-1) + 2f(n-2)$ using generating functions, set

$$G(z) = \sum_{n=1}^{\infty} f(n) z^n.$$

The recursion yields

$$G(z) = z + 3z^2 + \sum_{n=3}^{\infty} f(n-1) z^n + 2 \sum_{n=3}^{\infty} f(n-2) z^n$$

$$= z + 3z^2 + z\big(G(z) - z\big) + 2z^2 G(z),$$

from which we find

$$G(z) = \frac{z + 2z^2}{1 - z - 2z^2} = z \left(\frac{4/3}{1 - 2z} - \frac{1/3}{1 + z} \right).$$

Expanding $(1 - 2z)^{-1}$ and $(1 + z)^{-1}$ into geometric series, we obtain

$$G(z) = \sum_{n=1}^{\infty} \frac{2^{n+1} + (-1)^n}{3} z^n,$$

and by equating coefficients we find

$$f(n) = \frac{2^{n+1} + (-1)^n}{3}.$$

Problem 111

Let $p(x, y)$ be a real polynomial.
a. If $p(x, y) = 0$ for infinitely many (x, y) on the unit circle $x^2 + y^2 = 1$, must $p(x, y) = 0$ on the unit circle?
b. If $p(x, y) = 0$ on the unit circle, is $p(x, y)$ necessarily divisible by $x^2 + y^2 - 1$?

Answer. "Yes" is the answer to both questions.

Solution 1.   a.   The unit circle minus the point $(-1, 0)$ can be parametrized by

$$x = \frac{1 - t^2}{1 + t^2}, \qquad y = \frac{2t}{1 + t^2}, \qquad -\infty < t < \infty.$$

(This parametrization is reminiscent of Pythagorean triples $m^2 - n^2, 2mn, m^2 + n^2$.) If the total degree (in x and y) of $p(x, y)$ is d, then

$$(1 + t^2)^d p \left(\frac{1 - t^2}{1 + t^2}, \frac{2t}{1 + t^2} \right)$$

is a polynomial in t. By our assumption, this polynomial has infinitely many zeros, hence it is identically zero. But then $p(x, y)$ is identically zero on the unit circle minus the point $(-1, 0)$, and continuity forces $p(x, y)$ to be zero at $(-1, 0)$ as well.

   b.   View $p(x, y)$ as a polynomial in x (whose coefficients are polynomials in y). Using long division to divide $p(x, y)$ by $x^2 + y^2 - 1$, we can write

$$p(x, y) = q(x, y)(x^2 + y^2 - 1) + r(x, y),$$

where the degree in x of the remainder $r(x, y)$ is less than 2. We can then write $r(x, y) = f(y)x + g(y)$ for some polynomials $f(y)$ and $g(y)$. Since $p(x, y)$ and

$x^2 + y^2 - 1$ are both zero on the unit circle, so is $f(y)x + g(y)$. Furthermore, if (x, y) is on the unit circle, $(-x, y)$ is also, and so we have

$$f(y)(-x) + g(y) = f(y)x + g(y) = 0.$$

Therefore, $f(y) = g(y) = 0$ for all $y \in (-1, 1)$, and thus the polynomials $f(y)$ and $g(y)$ are identically zero. That is, $p(x, y) = q(x, y)(x^2 + y^2 - 1)$, and we are done.

Solution 2. We get the answers to (a) and (b) simultaneously, by showing:

c. If $p(x, y) = 0$ for infinitely many (x, y) on the unit circle, then $p(x, y)$ is divisible by $x^2 + y^2 - 1$.

This clearly answers (b) affirmatively; it also answers (a), because any polynomial which is divisible by $x^2 + y^2 - 1$ is identically zero on the unit circle.

Suppose $p(x, y) = 0$ for infinitely many (x, y) on the unit circle. As in Solution 1, we use long division to write

$$p(x, y) = q(x, y)(x^2 + y^2 - 1) + r(x, y), \qquad r(x, y) = f(y)x + g(y),$$

for some polynomials $f(y)$ and $g(y)$, and it is enough to show that $f(y)$ and $g(y)$ are identically zero.

For the infinitely many points on the unit circle where $p(x, y) = 0$, we have both $x^2 + y^2 - 1 = 0$ and $f(y)x + g(y) = 0$, and we can eliminate x to get

$$\left(g(y)\right)^2 + y^2 \left(f(y)\right)^2 - \left(f(y)\right)^2 = 0.$$

Therefore, the polynomial

$$\left(g(y)\right)^2 + y^2 \left(f(y)\right)^2 - \left(f(y)\right)^2$$

is zero for infinitely many values of y, so it is identically zero. That is,

$$\left(g(y)\right)^2 + y^2 \left(f(y)\right)^2 = \left(f(y)\right)^2$$

as polynomials in y.

Note that because the leading coefficients of $(g(y))^2$ and $y^2(f(y))^2$ are both nonnegative, they cannot cancel when we add these squares together. Therefore, the degree of $(f(y))^2$ is at least the degree of $y^2(f(y))^2$. However, if $f(y)$ is not identically zero, the degree of $y^2(f(y))^2$ is two more than the degree of $(f(y))^2$, a contradiction. So $f(y)$, and hence also $g(y)$, is identically zero, and we are done.

Problem 112

Find all real polynomials $p(x)$, whose roots are real, for which

$$p(x^2 - 1) = p(x)p(-x).$$

Idea. If α is a root of $p(x)$, then $\alpha^2 - 1$ is also, and yet $p(x)$ can have only finitely many roots.

Solution. The desired polynomials are 0 and those of the form

$$p_{j,k,l}(x) = (x^2 + x)^j (\varphi - x)^k (\overline{\varphi} - x)^l,$$

where $\varphi = (1 + \sqrt{5})/2$, $\overline{\varphi} = (1 - \sqrt{5})/2$ are the roots of $x^2 - x - 1 = 0$ and $j, k, l \geq 0$ are integers.

It is straightforward to check that the $p_{j,k,l}(x)$ do satisfy the functional equation $p(x^2 - 1) = p(x) p(-x)$. To see how they were found, observe that if α is a root of $p(x)$, then $p(\alpha^2 - 1) = p(\alpha)p(-\alpha) = 0$, so $\alpha^2 - 1$ is also a root of $p(x)$. But then $(\alpha^2 - 1)^2 - 1$ is, in turn, a root; more generally, each term of the iterative sequence

$$\alpha, \quad \alpha^2 - 1, \quad (\alpha^2 - 1)^2 - 1, \quad \ldots \tag{$*$}$$

is a root of $p(x)$. Since $p(x)$ has only finitely many roots, $(*)$ must eventually become periodic. The easiest way for this to happen is to have $\alpha^2 - 1 = \alpha$ (period 1), which yields $\alpha = \varphi$ or $\alpha = \overline{\varphi}$. In this case $p(x)$ is divisible by $\varphi - x$ or $\overline{\varphi} - x$, respectively.

The next case to consider is $(\alpha^2 - 1)^2 - 1 = \alpha$. Besides φ and $\overline{\varphi}$, this fourth-degree equation for α has roots 0 and -1. Note that if either 0 or -1 is a root of $p(x)$, so is the other, because $0^2 - 1 = -1$ and $(-1)^2 - 1 = 0$. So in this case $p(x)$ is divisible by $x^2 + x$.

We now show that it is not necessary to study any further cases, that is, that every nonzero polynomial $p(x)$ satisfying the given functional equation is one of the $p_{j,k,l}(x)$. Note that if $p(x)$ satisfies the given functional equation and is divisible by $p_{j,k,l}(x)$, then the polynomial $p(x)/p_{j,k,l}(x)$ also satisfies the functional equation. So we can divide out by any factors $\varphi - x$, $\overline{\varphi} - x$, and $x^2 + x$ that $p(x)$ may have, and assume that $p(x)$ does not have any of $-1, 0, \varphi, \overline{\varphi}$ as roots.

If $p(x)$ is nonconstant, let α_0 be the smallest root of $p(x)$. If $\alpha_0 > \varphi$, then $\alpha_0^2 - 1 > \alpha_0$, and the sequence $(*)$ will be strictly increasing, yielding infinitely many roots, a contradiction. If $\overline{\varphi} < \alpha_0 < \varphi$, then $\alpha_0^2 - 1 < \alpha_0$, contradicting our choice of α_0. Thus, we must have $\alpha_0 < \overline{\varphi}$. However, since α_0 is a root of $p(x)$, $p(x)$ has a factor $x - \alpha_0$, and so $p(x^2 - 1)$ has a factor $x^2 - 1 - \alpha_0$. Since

$p(x)$ has real roots, $p(x^2 - 1) = p(x)p(-x)$ factors into linear factors, hence $1 + \alpha_0 \geq 0$, and we have $-1 < \alpha_0 < \overline{\varphi}$.

The third term in the sequence $(*)$ is

$$(\alpha_0^2 - 1)^2 - 1 = \alpha_0 + ((\alpha_0^2 - 1)^2 - 1 - \alpha_0) = \alpha_0 + \alpha_0(\alpha_0 + 1)(\alpha_0 - \varphi)(\alpha_0 - \overline{\varphi}).$$

Because $-1 < \alpha_0 < \overline{\varphi}$, the product on the right is negative, which contradicts our choice of α_0. We conclude that $p(x)$ is constant. The functional equation now shows that $p(x) = 1 = p_{0,0,0}(x)$, and we are done.

Problem 113

Consider sequences of points in the plane that are obtained as follows: The first point of each sequence is the origin. The second point is reached from the first by moving one unit in any of the four "axis" directions (east, north, west, south). The third point is reached from the second by moving $1/2$ unit in any of the four axis directions (but not necessarily in the same direction), and so on. We call a point *approachable* if it is the limit of some sequence of the above type. Describe the set of all approachable points in the plane.

Solution. The set of all approachable points in the plane is the set of all points (x, y) for which $|x| + |y| \leq 2$ (see Figure 52).

First we show that $|x| + |y| \leq 2$ for every approachable point (x, y). Let (x, y) be approachable, and let $(x_0, y_0) = (0, 0), (x_1, y_1), \ldots, (x_n, y_n), \ldots$ be a sequence, of the type described in the problem, whose limit is (x, y). As we

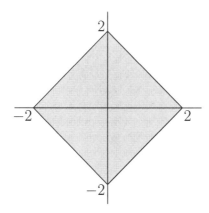

FIGURE 52

move from one point in the sequence to the next, either the x-coordinate or the y-coordinate changes by $1, \frac{1}{2}, \frac{1}{4}, \ldots$, while the other coordinate does not change. Thus, after n steps we will have

$$|x_n| + |y_n| \leq 1 + \frac{1}{2} + \cdots + \frac{1}{2^n} = 2 - \frac{1}{2^n}.$$

Since (x_n, y_n) approaches (x, y) as $n \to \infty$, it follows that $|x| + |y| \leq 2$.

To show that every point (x, y) with $|x| + |y| \leq 2$ is approachable, begin by dividing the set of such points into four congruent parts, as shown in Figure 53.

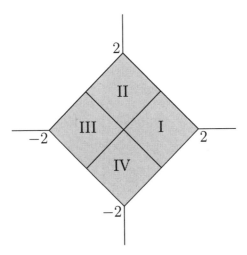

FIGURE 53

By symmetry, it is enough to show that any point in part I is approachable. To approach a point in part I, take the first "step" (move) to the right. We then have $(x_0, y_0) = (0, 0)$, $(x_1, y_1) = (1, 0)$. Now we have to show that we can approach the point (x, y) from the point $(1, 0)$ by moves of $\frac{1}{2}, \frac{1}{4}, \ldots$ in the axis directions. But this is exactly the original problem scaled down by a factor 2: we have a square whose sides are half the sides of the original square, and we are again at the center of that square. So to decide where to go next, we subdivide square I into four equal squares, as shown in Figure 54, and then move right, up, left, or down, depending on whether (x, y) is in A, B, C, or D.

Continuing this process, we obtain a sequence of points (x_n, y_n). To show that the limit of this sequence is (x, y), note that the distance from (x_n, y_n) to

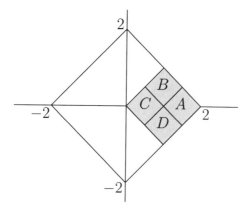

FIGURE 54

(x, y) is at most half the length of a diagonal of the nth new square. Since this length is halved at each stage, $\lim_{n \to \infty} (x_n, y_n) = (x, y)$, and we are done.

Problem 114

A gambling game is played as follows: D dollar bills are distributed in some manner among N indistinguishable envelopes, which are then mixed up in a large bag. The player buys random envelopes, one at a time, for one dollar and examines their contents as they are purchased. If the player can buy as many or as few envelopes as desired, and, furthermore, knows the initial distribution of the money, then for what distribution(s) will the player's expected net return be maximized?

Solution. If $D > (N + 1)/2$, the player's expected net return will be maximized if and only if all D dollars are in a single envelope; if $D \le (N + 1)/2$, the distribution does not matter.

If, at some stage in the game, it is in the player's best interest to buy an envelope, then he or she should certainly continue buying envelopes until one of them is found to contain money. (At that point, the player may or may not prefer to stop buying.) In particular, if all D dollar bills are in the same envelope, then the player will either not buy at all or buy envelopes until the one containing the money is purchased. If the latter strategy is followed, the envelope with the money is equally likely to be bought first, second, ..., or

Nth, and so the player's expected investment to get the D dollars will be

$$\sum_{j=1}^{N} j \cdot \frac{1}{N} = \frac{N+1}{2}.$$

Therefore, still assuming all the bills are in one envelope, the player will buy envelopes if $D > (N+1)/2$, with an expected net return of $D - (N+1)/2$, and the player will not bother to play at all if $D \leq (N+1)/2$.

Now we will show that from the player's point of view, any other distribution of the bills among the envelopes is worse if $D > (N+1)/2$ and no better (that is, there is still no reason to play) if $D \leq (N+1)/2$.

To see why, let $E(D, N)$ denote the maximum net return, over all possible distributions, when D bills are distributed over N envelopes. (This maximum exists because the number of distributions is finite, as is the number of possible strategies for the player given a distribution.)

Note that $E(D, N - 1) \geq E(D, N)$, because merging the contents of two envelopes and throwing away the now empty envelope can only help the player.

Our claim is that

$$E(D, N) = \begin{cases} D - (N+1)/2 & \text{if } D > (N+1)/2 \\ 0 & \text{if } D \leq (N+1)/2, \end{cases}$$

and that in the case $D > (N+1)/2$, the maximum *only* occurs when all the bills are in the same envelope.

We prove this by induction on N. If $N = 1$, there is only one possible distribution, and the claim follows. Suppose the claim is correct for $N - 1$ and all values of D. If $E(D, N) = 0$, then we must have $D \leq (N+1)/2$ (otherwise, we could do better by putting all bills in the same envelope), and we are done. So assume $E(D, N) > 0$. Suppose the maximum occurs when M of the N envelopes contain money, say $D_1, D_2, \cdots, D_M$ dollars respectively, with $D = \sum_{m=1}^{M} D_m$. Then after the player buys the first envelope, which she or he will do because $E(D, N) > 0$, the player will have lost one dollar with probability $(N - M)/N$ and, for $1 \leq m \leq M$, gained $D_m - 1$ dollars with probability $1/N$. We therefore have the inequality

$$E(D, N) \leq \frac{N - M}{N} \left(E(D, N - 1) - 1 \right)$$

$$+ \sum_{m=1}^{M} \frac{1}{N} \left(E(D - D_m, N - 1) + D_m - 1 \right)$$

$$= \frac{N - M}{N} E(D, N - 1) + \frac{1}{N} \sum_{m=1}^{M} E(D - D_m, N - 1) + \frac{D}{N} - 1. \quad (*)$$

Since $E(D, N-1) \geq E(D, N) > 0$, we know from the induction hypothesis that $E(D, N-1) = D - N/2$. On the other hand, it is not clear whether $E(D - D_m, N-1) > 0$, and so we distinguish two cases.

Case 1. $E(D - D_m, N-1) > 0$ for all m.
    In this case inequality $(*)$, together with the induction hypothesis, yields

$$E(D, N) \leq \frac{N-M}{N}\left(D - \frac{N}{2}\right) + \frac{1}{N}\sum_{m=1}^{M}\left(D - D_m - \frac{N}{2}\right) + \frac{D}{N} - 1$$

$$= D - \frac{N}{2} - 1$$

$$< D - \frac{N+1}{2},$$

and we are done.

Case 2. $E(D - D_m, N-1) = 0$ for at least one m.
    Since $E(D - D_m, N-1) < E(D, N-1)$ for all m, we then have

$$\sum_{m=1}^{M} E(D - D_m, N-1) \leq (M-1)E(D, N-1),$$

with equality only for $M = 1$. Inequality $(*)$ then yields

$$E(D, N) \leq \frac{N-M}{N}E(D, N-1) + \frac{1}{N}(M-1)E(D, N-1) + \frac{D}{N} - 1$$

$$= \frac{N-1}{N}E(D, N-1) + \frac{D}{N} - 1$$

$$= \frac{N-1}{N}\left(D - \frac{N}{2}\right) + \frac{D}{N} - 1$$

$$= D - \frac{N+1}{2},$$

with equality only for $M = 1$. That is the situation in which all bills are in one envelope, and we are done.

Problem 115

Let $\alpha = .d_1 d_2 d_3 \ldots$ be a decimal representation of a real number between 0 and 1. Let r be a real number with $|r| < 1$.
a. If α and r are rational, must $\sum_{i=1}^{\infty} d_i r^i$ be rational?

b. If α and r are rational, must $\sum_{i=1}^{\infty} id_i r^i$ be rational?

c. If r and $\sum_{i=1}^{\infty} d_i r^i$ are rational, must α be rational?

Answer. "Yes" to (a) and (b); "no" to (c).

Solution 1. a. Suppose that α is rational. Then the decimal representation is eventually periodic; let p be its period, and let k be the number of digits before the periodic behavior starts. We have

$$\alpha = .d_1 d_2 \ldots d_k \underbrace{d_{k+1} \ldots d_{k+p}} \underbrace{d_{k+1} \ldots d_{k+p}} \underbrace{d_{k+1} \ldots d_{k+p}} \ldots .$$

For any real number r with $|r| < 1$, we can define

$$x = \sum_{i=1}^{k} d_i r^i \quad \text{and} \quad y = \sum_{i=1}^{p} d_{k+i} r^{k+i},$$

and we then have

$$\sum_{i=1}^{\infty} d_i r^i = x + y + y \, r^p + y \, (r^p)^2 + y \, (r^p)^3 + \cdots .$$

Summing the infinite geometric series, we obtain

$$\sum_{i=1}^{\infty} d_i r^i = x + \frac{y}{1 - r^p}. \tag{$*$}$$

Now if r is rational, then so are x and y. But then $\sum_{i=1}^{\infty} d_i r^i$ is also rational, and we are done.

b. Again, suppose that α is rational. Note that the sum $\sum_{i=1}^{\infty} id_i r^i$ can be rewritten as $r \sum_{i=1}^{\infty} id_i r^{i-1}$; the new sum looks like a derivative.

In fact, we can get the new sum by differentiating both sides of the identity $(*)$, which holds for all r with $|r| < 1$. If we then multiply by r, we find

$$\sum_{i=1}^{\infty} id_i r^i = \sum_{i=1}^{k} id_i r^i + \left(\frac{pr^p}{(1 - r^p)^2} \right) \sum_{i=1}^{p} d_{k+i} r^{k+i}$$

$$+ \left(\frac{1}{1 - r^p} \right) \sum_{i=1}^{p} (k + i) d_{k+i} r^{k+i}.$$

If r is a rational number, then the three finite sums on the right are rational; therefore, $\sum_{i=1}^{\infty} id_i r^i$ is also rational, and we are done.

c. If r has the form $1/b$ where $b \geq 10$ is an integer, then α is rational. For in that case, if $\sum_{i=1}^{\infty} d_i b^{-i}$ is rational, then $.d_1 d_2 d_3 \ldots$ is the base b expansion of a rational number, hence the sequence $d_1, d_2, d_3, \ldots$ is eventually periodic, and α is also rational.

If r does not have this form, however, α may fail to be rational. To see this, consider the case $r = 1/2$ and the sum

$$\sum_{i=1}^{\infty} \left(\frac{1}{2}\right)^{2i} = \left(\frac{1}{4}\right) + \left(\frac{1}{4}\right)^2 + \left(\frac{1}{4}\right)^3 + \cdots = \frac{1/4}{1-(1/4)} = \frac{1}{3}.$$

This sum can be written as $\sum_{i=1}^{\infty} d_i r^i$ with $d_{2i} = 1$, $d_{2i+1} = 0$ for all i. But we can also, independently for each $i = 1, 2, \ldots$, choose (d_{2i}, d_{2i+1}) to be either $(1, 0)$ or $(0, 2)$; regardless of our choices, if we again set $d_1 = 0$, we will have

$$\sum_{i=1}^{\infty} d_i r^i = \sum_{i=1}^{\infty} \left(d_{2i} \left(\frac{1}{2}\right)^{2i} + d_{2i+1} \left(\frac{1}{2}\right)^{2i+1} \right) = \sum_{i=1}^{\infty} \left(\frac{1}{2}\right)^{2i} = \frac{1}{3}.$$

Furthermore, we can make these choices in such a way that the resulting decimal representation $\alpha = .d_1 d_2 d_3 \ldots$ will be nonperiodic; as a result, α will be irrational. (For example, choose $(d_{2i}, d_{2i+1}) = (0, 2)$ precisely when i is a power of 2.)

Solution 2.   a.   As in Solution 1, let p be the (eventual) period of the decimal representation $.d_1 d_2 d_3 \ldots$ of α, and suppose that the periodic behavior begins with d_{k+1}, so that $d_{i+p} = d_i$ for all $i > k$. Now recall the technique, which is often used to find the fractional representation of a rational number given its periodic decimal expansion, of "shifting over" an expansion and then subtracting it from the original. Using this technique, we find

$$(1 - r^p) \sum_{i=1}^{\infty} d_i r^i = \sum_{i=1}^{p} d_i r^i + \sum_{i=p+1}^{p+k} (d_i - d_{i-p}) r^i + \sum_{i=p+k+1}^{\infty} (d_i - d_{i-p}) r^i$$

$$= \sum_{i=1}^{p} d_i r^i + \sum_{i=p+1}^{p+k} (d_i - d_{i-p}) r^i$$

for $|r| < 1$. This final expression is a finite sum, so if r is rational, then $\sum_{i=1}^{\infty} d_i r^i$ is also rational.

b. Using the same idea as in (a), we have

$$(1 - r^p)\sum_{i=1}^{\infty} id_i r^i = \sum_{i=1}^{p} id_i r^i + \sum_{i=p+1}^{p+k} (id_i - (i-p)d_{i-p})r^i$$

$$+ \sum_{i=p+k+1}^{\infty} (id_i - (i-p)d_{i-p})r^i$$

$$= \sum_{i=1}^{p} id_i r^i + \sum_{i=p+1}^{p+k} (id_i - (i-p)d_{i-p})r^i + \sum_{i=p+k+1}^{\infty} pd_{i-p}r^i.$$

If r is rational, the first two (finite) sums on the right are certainly rational, and the third sum, which may be rewritten as $pr^{p+k}\sum_{i=1}^{\infty} d_{i+k}r^i$, is rational by the result of (a). Therefore, $\sum_{i=1}^{\infty} id_i r^i$ is rational.

c. Let $r = 1/2$, and choose $d_1 = 0$ and (d_{2i}, d_{2i+1}) just as in Solution 1. Now observe that there are uncountably many such choices for the sequence $d_1, d_2, d_3, \ldots$, no two of which yield the same α. Since the rationals are countable, some ("most") of the possible α's are irrational. (In fact, since the set of algebraic numbers is countable, the α's are generally transcendental.)

Problem 116

Let $\mathcal{L}_1$ and $\mathcal{L}_2$ be skew lines in space (that is, straight lines which do not lie in the same plane). How many straight lines $\mathcal{L}$ have the property that every point on $\mathcal{L}$ has the same distance to $\mathcal{L}_1$ as to $\mathcal{L}_2$?

Solution.   There are infinitely many lines $\mathcal{L}$ such that every point on $\mathcal{L}$ has the same distance to $\mathcal{L}_1$ as to $\mathcal{L}_2$. To show this, we will first choose convenient coordinates in $\mathbf{R}^3$, then find the set of all points equidistant from $\mathcal{L}_1$ and $\mathcal{L}_2$, and finally find all lines contained in this set of points.

We can set up our coordinate system so that the points on $\mathcal{L}_1$ and $\mathcal{L}_2$ which are closest to each other are $(0, 0, 1)$ and $(0, 0, -1)$. In this case, $\mathcal{L}_1$ lies in the plane $z = 1$ and $\mathcal{L}_2$ lies in the plane $z = -1$. After a suitable rotation about the z-axis, we may assume that for some $m > 0$, $\mathcal{L}_1$ is described by $y = mx$, $z = 1$, while $\mathcal{L}_2$ is given by $y = -mx$, $z = -1$.

The point $(t, mt, 1)$ on $\mathcal{L}_1$ closest to a given point (x, y, z) can be found by minimizing the square of the distance,

$$(t - x)^2 + (mt - y)^2 + (1 - z)^2.$$

A short computation shows that the minimum occurs for $t = (x+my)/(1+m^2)$. Therefore, the square of the distance from (x, y, z) to $\mathcal{L}_1$ is

$$\left(\frac{x+my}{1+m^2} - x\right)^2 + \left(\frac{mx+m^2y}{1+m^2} - y\right)^2 + (1-z)^2 = \frac{(y-mx)^2}{1+m^2} + (z-1)^2.$$

In a similar way, or by using the symmetry of the problem, we find that the square of the distance from (x, y, z) to $\mathcal{L}_2$ is

$$\frac{(y+mx)^2}{1+m^2} + (z+1)^2.$$

The point (x, y, z) is equidistant from $\mathcal{L}_1$ and $\mathcal{L}_2$ if and only if the two expressions above are equal. Simplifying, we find that the points (x, y, z) equidistant from $\mathcal{L}_1$ and $\mathcal{L}_2$ form the surface

$$z = -\frac{mxy}{1+m^2}.$$

We can write the equation of a line through a fixed point (x_0, y_0, z_0) on this surface in the form

$$(x_0 + at, y_0 + bt, z_0 + ct),$$

where (a, b, c) represents the direction of the line. For this line to lie entirely on the surface, we must have

$$z_0 + ct = -\frac{m(x_0 + at)(y_0 + bt)}{1+m^2}$$

for all t. Comparing coefficients of 1, t, and t^2, we see that this comes down to

$$z_0 = -\frac{mx_0y_0}{1+m^2}, \quad c = -\frac{m(bx_0 + ay_0)}{1+m^2}, \quad 0 = -\frac{mab}{1+m^2}.$$

The first equation is automatically satisfied, because (x_0, y_0, z_0) is on the surface. The last equation implies $a = 0$ or $b = 0$, so up to constant multiples there are two solutions to the equations: $(a, b, c) = (0, 1, -mx_0/(1+m^2))$ and $(a, b, c) = (1, 0, -my_0/(1 + m^2))$. Thus, through the point (x_0, y_0, z_0) on the surface there are exactly two lines $\mathcal{L}$ such that every point on $\mathcal{L}$ has the same distance to $\mathcal{L}_1$ as to $\mathcal{L}_2$. Since x_0 and y_0 can be varied at will, there are, in all, infinitely many such lines.

Problem 117

We call a sequence $(x_n)_{n\geq 1}$ a *superinteger* if (i) each x_n is a nonnegative integer less than 10^n and (ii) the last n digits of x_{n+1} form x_n. One example of such a sequence is $1, 21, 021, 1021, 21021, 021021, \ldots$, which we abbreviate by $\ldots 21021$. We can do arithmetic with superintegers; for instance, if x is the superinteger above, then the product xy of x with the superinteger $y = \ldots 66666$ is found as follows:

$1 \times 6 = 6$: the last digit of xy is 6.

$21 \times 66 = 1386$: the last two digits of xy are 86.

$021 \times 666 = 13986$: the last three digits of xy are 986.

$1021 \times 6666 = 6805986$: the last four digits of xy are 5986, etc.

Is it possible for two nonzero superintegers to have product $0 = \ldots 00000$?

Solution. Yes, it is possible for two nonzero superintegers to have product zero.

In fact, we will show that there exist nonzero superintegers $x = \ldots a_n \ldots a_2 a_1$ and $y = \ldots b_n \ldots b_2 b_1$ such that for any $k \geq 1$, 2^k divides $a_k \ldots a_2 a_1$ and 5^k divides $b_k \ldots b_2 b_1$. Then 10^k divides $a_k \ldots a_2 a_1 \times b_k \ldots b_2 b_1$, so the product xy ends in k zeros for any k; that is, xy is zero, and we are done.

We start with x, and we show by induction that there exist digits $a_1, a_2, \ldots$ such that for all $k \geq 1$, 2^k divides $a_k \ldots a_2 a_1$. Take $a_1 = 2$ to show this is true for $k = 1$. Now suppose $a_1, a_2, \ldots, a_m$ have been found such that 2^m divides $a_m \ldots a_2 a_1$, so $c_m = (a_m \ldots a_2 a_1)/2^m$ is an integer. We want to find a digit $d = a_{m+1}$ for which 2^{m+1} divides $d \cdot 10^m + a_m \ldots a_2 a_1$, that is, for which 2 divides $d \cdot 5^m + c_m$. Since 5^m is odd, we can take $d = 0$ if c_m is even and $d = 1$ if c_m is odd. (The actual superinteger x constructed by this inductive procedure ends in $\ldots 010112$.)

Similarly, we show that y exists by starting with $b_1 = 5$ and showing that there exist digits $b_2, b_3, \ldots$ such that for all $k \geq 1$, 5^k divides $b_k \ldots b_2 b_1$. This time, assuming 5^m divides $b_m \ldots b_2 b_1$, we want to find a digit d such that 5^{m+1} divides $d \cdot 10^m + b_m \ldots b_2 b_1$, that is, such that 5 divides $d \cdot 2^m + (b_m \ldots b_2 b_1)/5^m$. In other words, we want a solution to $2^m d \equiv -(b_m \ldots b_2 b_1)/5^m \pmod 5$. Since 2^m has an inverse modulo 5, exactly one of $0, 1, 2, 3, 4$ is a solution for d, and we are done. (The actual superinteger y constructed in this way ends in $\ldots 203125$.)

Problem 118

If $\sum a_n$ converges, must there exist a periodic function $\varepsilon : \mathbf{Z} \to \{1, -1\}$ such that $\sum \varepsilon(n) |a_n|$ converges?

Idea. Construct a series consisting of a few "large" positive terms just balanced by many "small" negative terms. To prevent cancellation among the large terms in $\sum \varepsilon(n)|a_n|$, create gaps of $1!, 2!, 3!, \ldots$ between them.

Solution. No, there exist convergent series $\sum a_n$ such that $\sum \varepsilon(n)|a_n|$ diverges for every periodic function $\varepsilon : \mathbf{Z} \to \{1, -1\}$.

We will show that one such series is given by

$$
a_n = \begin{cases}
\dfrac{1}{\ln k} & \text{if } n = \sum_{j=2}^{k} j! \text{ for some } k \geq 2, \\[2ex]
-\dfrac{1}{k!\ln k} & \text{if } \sum_{j=2}^{k-1} j! < n < \sum_{j=2}^{k} j! \text{ for some } k \geq 3, \\[2ex]
-\dfrac{1}{2\ln 2} & \text{if } n = 1.
\end{cases}
$$

Let $s_k = \sum_{j=2}^{k} j!$. Then for $s_{k-1} < n \leq s_k$, $k \geq 3$, we have

$$
-\frac{1}{\ln k} + \sum_{j=2}^{k} \frac{1}{j!\ln j} < \sum_{j=1}^{n} a_j \leq \sum_{j=2}^{k} \frac{1}{j!\ln j},
$$

from which it follows that $\sum a_n$ converges.

Now let $\varepsilon : \mathbf{Z} \to \{1, -1\}$ be periodic, say with period N; we must show that $\sum \varepsilon(n)|a_n|$ diverges. First consider the "large" terms

$$
\varepsilon(s_k)|a_{s_k}| = \pm\frac{1}{\ln k}.
$$

We know that these terms will have the same sign for all $k \geq \max(N-1, 2)$, since for those k, $s_{k+1} - s_k$ is a multiple of N. By replacing ε by $-\varepsilon$ if necessary, we may assume that this sign is positive. As for the "small" terms, at least every Nth one must be positive, because of the periodicity of ε. So, provided $k \geq k_0 = \max(N, 3)$, among the $k!$ terms $\varepsilon(n)|a_n|$ with $s_{k-1} < n \leq s_k$, one (for $n = s_k$) will be $\dfrac{1}{\ln k}$, $\dfrac{k!}{N} - 1$ others will also be positive, and the remaining $((N-1)/N)k!$ terms might be either positive or negative. Therefore, we have

$$
\sum_{n=s_{k-1}+1}^{s_k} \varepsilon(n)|a_n| \geq \frac{1}{\ln k} + \left(\frac{k!}{N} - 1\right)\frac{1}{k!\ln k} - \left(\frac{N-1}{N}k!\right)\frac{1}{k!\ln k}
$$

$$
= \left(\frac{2}{N} - \frac{1}{k!}\right)\frac{1}{\ln k} > \frac{1}{N\ln k}.
$$

If we add these inequalities for $k = k_0, k_0 + 1, \ldots$, we see that

$$
\sum_{n=s_{k_0-1}+1}^{\infty} \varepsilon(n)|a_n| > \frac{1}{N} \sum_{k=k_0}^{\infty} \frac{1}{\ln k}.
$$

Since $\sum \dfrac{1}{\ln k}$ diverges, $\sum \varepsilon(n)|a_n|$ must also diverge, and we are done.

Problem 119

Let $f(x) = x - 1/x$. For any real number x_0, consider the sequence defined by x_0, $x_1 = f(x_0)$, ... , $x_{n+1} = f(x_n)$, ..., provided $x_n \neq 0$. Define x_0 to be a *T-number* if the sequence terminates, that is, if $x_n = 0$ for some n.

a. Show that the set of all T-numbers is countably infinite (denumerable).

b. Does every open interval contain a T-number?

Solution. a. Let

$$S_n = \{x_0 \in \mathbf{R} \mid x_n = \underbrace{f(f(\ldots(f(x_0))\ldots))}_{n} = 0\}.$$

We prove by induction that S_n has exactly 2^n elements. This is true for $n = 0$, since 0 is the only element of S_0; assume it is true for n. Note that $x_0 \in S_{n+1}$ if and only if $f(x_0) \in S_n$. Also, the equality $f(x_0) = x_0 - 1/x_0$ can be rewritten as $x_0^2 - f(x_0)x_0 - 1 = 0$, and, given $f(x_0)$, this quadratic equation for x_0 has two real roots. Since there are 2^n possibilities for $f(x_0) \in S_n$, there are $2 \cdot 2^n = 2^{n+1}$ elements $x_0 \in S_{n+1}$, and the induction step is complete.

Now since the set of all T-numbers can be written as $\bigcup_{n=0}^{\infty} S_n$, a countable union of finite sets, it is countable, and since, for any n, it has more than 2^n elements, the set is infinite.

b. Yes, every open interval does contain a T-number. Suppose that, on the contrary, there exist open intervals which do not contain any T-numbers; we will call such intervals *T-free*. Note that since 0 is a T-number, any T-free interval (a, b) satisfies either $0 \leq a < b$ or $a < b \leq 0$. Also, since f is an odd function, the set of all T-numbers is symmetric about 0, so for any T-free interval (a, b), the interval $(-b, -a)$ is also T-free. Thus there exists some T-free interval (a, b) with $0 < a < b$ (if $a = 0$, shrink the interval slightly). We will show the following:

I. For every T-free interval (a, b) with $0 < a < b$, we can find a T–free interval (c, d) with $0 < c < d \leq 1$ which is *at least* as long as (a, b).

II. For every T-free interval (a, b) with $0 < a < b \leq 1$, we can find a T-free interval (c, d) with $0 < c < d$ which is more than *twice* as long as (a, b).

Repeatedly combining (I) and (II) then yields arbitrarily long T-free intervals which are contained in $(0, 1)$, a contradiction.

Let (a, b) be any T-free interval with $0 < a < b$. Note that f is increasing and continuous on $(0, \infty)$; therefore, the image of (a, b) under f is the interval $(f(a), f(b))$. Since the image of a T-free set is necessarily T-free, this new interval is again T-free, and since its length is

$$f(b) - f(a) = (b - a) + (1/a - 1/b) > b - a,$$

it is longer than the original interval.

To prove (I) above, note that if $b \leq 1$, we can take $(c, d) = (a, b)$. If $b > 1$, consider the successive T-free intervals $\big(f(a), f(b)\big)$, $\big(f(f(a)), f(f(b))\big)$, and so forth. We have $f(b) = b - 1/b$, $f\big(f(b)\big) = b - 1/b - 1/f(b) < b - 2/b$, and continuing in this way,

$$\underbrace{f\big(f(\cdots f(b) \cdots)\big)}_{n} \leq b - \frac{n}{b}.$$

Thus for large enough n, we have

$$\underbrace{f\big(f(\cdots f(b) \cdots)\big)}_{n} \leq 1.$$

If n_0 is the first value of n for which this happens, then

$$\underbrace{f\big(f(\cdots f(b) \cdots)\big)}_{n_0} > 0,$$

since f maps any number greater than 1 to a positive number. We can then take

$$(c, d) = \left(\underbrace{f\big(f(\cdots f(a) \cdots)\big)}_{n_0} , \underbrace{f\big(f(\cdots f(b) \cdots)\big)}_{n_0} \right),$$

which is a T-free interval with $0 < d \leq 1$ and therefore $0 \leq c < d \leq 1$. By the above, this interval is longer than (a, b), so if $c = 0$, we can shrink the interval slightly; the proof of (I) is now complete.

To prove (II), we now assume $0 < a < b \leq 1$. Thus we have $f(b) \leq 0$. Therefore, using the symmetry mentioned earlier, $(c, d) = \big(-f(b), -f(a)\big)$ is a T-free interval with $0 \leq c < d$. Its length is

$$f(b) - f(a) = b - a + \frac{b - a}{ab} > 2(b - a),$$

and we are done. (Again, if $c = 0$, we can shrink the new interval slightly.)

Comment. For each n, x_n is a rational function of x_0 with rational coefficients, hence every T-number is an *algebraic number* (a root of a polynomial with rational coefficients). Since the set of all algebraic numbers is countable, we see again that the set of T-numbers is countable. However, this argument does not show that there are infinitely many T-numbers.

Problem 120

For n a positive integer, show that the number of integral solutions (x, y) of $x^2 + xy + y^2 = n$ is finite and a multiple of 6.

Idea. If (x, y) is an integral solution of $x^2 + xy + y^2 = n$, then $(-x, -y)$ is a different solution, so solutions come in pairs. If we can show instead that solutions come in sixes (and that there are only finitely many), we will be done. To see why solutions come in sixes, we can use algebraic manipulation to rewrite $x^2 + xy + y^2$ as $a^2 + ab + b^2$ for suitable $(a, b) \neq (x, y)$.

Solution. First note that for any solution (x, y), we have

$$2n = 2x^2 + 2xy + 2y^2 = x^2 + y^2 + (x + y)^2 \geq x^2 + y^2.$$

Therefore, any integral solution is one of the lattice points (points whose coordinates are integers) on or inside a circle of radius $\sqrt{2n}$, and so the number of integral solutions is finite.

Now observe that

$$x^2 + xy + y^2 = (x + y)^2 - xy$$
$$= (x + y)^2 - x(x + y) + x^2$$
$$= (x + y)^2 + (x + y)(-x) + (-x)^2.$$

Thus, if (x, y) is an integral solution of $x^2 + xy + y^2 = n$, then so is $(x+y, -x)$. If we repeat this process with the new solution, we go through a cycle of solutions:

$$(x, y), \ (x + y, -x), \ (y, -x - y), \ (-x, -y), \ (-x - y, x), \ (-y, x + y), \quad (*)$$

after which we get back to (x, y). It can be checked directly that, since x and y cannot both be zero, all six solutions in the cycle $(*)$ are different.

Alternatively, we can use a bit of linear algebra. Because

$$\begin{pmatrix} x + y \\ -x \end{pmatrix} = \begin{pmatrix} 1 & 1 \\ -1 & 0 \end{pmatrix} \begin{pmatrix} x \\ y \end{pmatrix},$$

the solutions in the cycle, rewritten as column vectors, are given by

$$\begin{pmatrix} 1 & 1 \\ -1 & 0 \end{pmatrix}^k \begin{pmatrix} x \\ y \end{pmatrix}$$

for $k = 0, 1, \ldots, 5$. The eigenvalues of the matrix $\mathbf{A} = \begin{pmatrix} 1 & 1 \\ -1 & 0 \end{pmatrix}$ are found to be $e^{2\pi i/6}$ and $e^{-2\pi i/6}$, so for $k = 1, 2, \ldots, 5$, $\mathbf{A}^k$ cannot have an eigenvalue 1, and hence $\mathbf{A}^k \begin{pmatrix} x \\ y \end{pmatrix}$ cannot equal $\begin{pmatrix} x \\ y \end{pmatrix}$.

Since all six solutions in $(*)$ are different, and since the set of all integral solutions of $x^2 + xy + y^2 = n$ can be partitioned into such cycles, the number of integral solutions is a multiple of 6, and we are done.

Comment. As pointed out by the late Ian Richards (University of Minnesota), another, perhaps more systematic, way to find the cycle $(*)$ starts by factoring $x^2 + xy + y^2$ into (complex) algebraic integers, as follows. Note that

$$x^3 - y^3 = (x - y)(x - \omega y)(x - \omega^2 y),$$

where $\omega \neq 1$ is a cube root of unity, and that for $x \neq y$,

$$x^2 + xy + y^2 = \frac{x^3 - y^3}{x - y}.$$

Therefore,

$$x^2 + xy + y^2 = (x - \omega y)(x - \omega^2 y);$$

this factorization actually holds whether or not $x = y$. Since $\omega^2 + \omega + 1 = 0$, the factorization can be rewritten as follows:

$$(x - \omega y)(x - \omega^2 y) = (x + y + \omega^2 y)(x + y + \omega y)$$

$$= (x + y + \omega y)(x + y + \omega^2 y)$$

$$= \big((x + y) - \omega(-y)\big)\big((x + y) - \omega^2(-y)\big).$$

Therefore, if (x, y) is a solution, so is $(x + y, -y)$. At first sight this is not too helpful, since repeating the process gets us right back to (x, y). However, by symmetry, once $(x+y, -y)$ is known to be a solution, so is $(-y, x+y)$. We now find the cycle $(*)$ in the "backward" direction: (x, y), $(-y, x + y)$, $(-x - y, x)$, and so forth.

Incidentally, the same symmetry $(x, y) \leftrightarrow (y, x)$ can be used to show that the number of integral solutions is often divisible by 12.

However, due to the fact that (y, x) may already be among the solutions in $(*)$, the number is not always divisible by 12. For instance, there are exactly six integral solutions for $n = 1$.

Problem 121

For what real numbers x can one say the following?

a. For each positive integer n, there exists an integer m such that

$$\left|x - \frac{m}{n}\right| < \frac{1}{3n}.$$

b. For each positive integer n, there exists an integer m such that

$$\left|x - \frac{m}{n}\right| \leq \frac{1}{3n}.$$

Idea. Consider those integers n which are powers of 2. If the approximations m/n to x for these n are sufficiently close to each other, they cannot be distinct.

Solution. a.   The condition holds for x if and only if x is an integer.

If x is an integer, then for any n we can take $m = xn$, so the condition holds for x. Conversely, suppose that the condition holds for x. Let m_k be the numerator corresponding to $n = 2^k$, $k = 0, 1, \ldots$. If $m_k/2^k \neq m_{k+1}/2^{k+1}$, then

$$\left|\frac{m_k}{2^k} - \frac{m_{k+1}}{2^{k+1}}\right| = \left|\frac{2m_k - m_{k+1}}{2^{k+1}}\right| \geq \frac{1}{2^{k+1}},$$

and so

$$\frac{1}{2^{k+1}} \leq \left|\frac{m_k}{2^k} - \frac{m_{k+1}}{2^{k+1}}\right| \leq \left|\frac{m_k}{2^k} - x\right| + \left|x - \frac{m_{k+1}}{2^{k+1}}\right|$$

$$< \frac{1}{3 \cdot 2^k} + \frac{1}{3 \cdot 2^{k+1}} = \frac{1}{2^{k+1}},$$

$$(*)$$

a contradiction. We conclude that $m_k/2^k = m_{k+1}/2^{k+1}$ for all k, so all the approximations $m_k/2^k$ to x are equal to the integer m_0. But then

$$|x - m_0| = \left|x - \frac{m_k}{2^k}\right| \leq \frac{1}{3 \cdot 2^k}$$

for all k; it follows that $x - m_0 = 0$, so x is an integer, and we are done.

b.   The condition holds for x if and only if $3x$ is an integer.

Suppose that $3x$ is an integer, and let n be a positive integer. Observe that $\left|x - \frac{m}{n}\right| \leq \frac{1}{3n}$ is equivalent to $|3nx - 3m| \leq 1$. Since the distance from the integer $3nx = n(3x)$ to the nearest integer multiple of 3 is at most 1, there exists an integer m for which $|3nx - 3m| \leq 1$; thus, the condition holds for x.

Now suppose that the condition holds for x. As in (a), let m_k be the numerator corresponding to $n = 2^k$. If $m_k/2^k = m_{k+1}/2^{k+1}$ for all k, then we see as in (a) that x is an integer, and we are done. Otherwise, there is an integer $K \geq 0$ such that

$$m_0 = \frac{m_1}{2} = \cdots = \frac{m_K}{2^K} \neq \frac{m_{K+1}}{2^{K+1}}.$$

For $k = K$, the inequalities in $(*)$ of (a) are equalities, and so

$$x = \frac{m_K}{2^K} \pm \frac{1}{3 \cdot 2^K} = m_0 \pm \frac{1}{3 \cdot 2^K}.$$

Now suppose that $K > 0$, and let $n = 3 \cdot 2^{K-1}$. Then for the corresponding numerator m, we have

$$\left| x - \frac{m}{3 \cdot 2^{K-1}} \right| \leq \frac{1}{9 \cdot 2^{K-1}}.$$

On the other hand, since $x = m_0 \pm 1/(3 \cdot 2^K)$, we have

$$x - \frac{m}{3 \cdot 2^{K-1}} = \frac{m_0 \cdot 3 \cdot 2^K \pm 1 - 2m}{3 \cdot 2^K}.$$

Since the numerator of this fraction is odd, $x - m/(3 \cdot 2^{K-1})$ is nonzero and

$$\left| x - \frac{m}{3 \cdot 2^{K-1}} \right| \geq \frac{1}{3 \cdot 2^K} > \frac{1}{9 \cdot 2^{K-1}},$$

a contradiction. We conclude that $K = 0$, $x = m_0 \pm \frac{1}{3}$, and we are done.

Problem 122

Let $\mathbf{Z}_n$ be the set $\{0, 1, \ldots, n-1\}$ with addition mod n. Consider subsets S_n of $\mathbf{Z}_n$ such that $(S_n + k) \cap S_n$ is nonempty for every k in $\mathbf{Z}_n$. Let $f(n)$ denote the minimal number of elements in such a subset. Find

$$\lim_{n \to \infty} \frac{\ln f(n)}{\ln n},$$

or show that this limit does not exist.

Solution. The limit is $1/2$.

First we rephrase the condition that $(S_n + k) \cap S_n$ is nonempty for all k, as follows: For every k in $\mathbf{Z}_n$, there are elements x and y in S_n such that $x - y \equiv k$ (mod n). We call S_n a *difference set* modulo n if this condition is satisfied.

For a difference set S_n with m elements, there are at most m^2 possible differences. This shows that $(f(n))^2 \geq n$, and therefore

$$\frac{\ln f(n)}{\ln n} \geq \frac{1}{2}.$$

On the other hand, let

$$T_n = \left\{1, 2, 3, \ldots, \lfloor\sqrt{n}\rfloor, 2\lfloor\sqrt{n}\rfloor, 3\lfloor\sqrt{n}\rfloor, \ldots, \lfloor\sqrt{n}\rfloor^2\right\}.$$

We claim that for $n \geq 16$, T_n is a difference set modulo n. Note that any integer from 0 to $\lfloor\sqrt{n}\rfloor^2 - 1$, inclusive, is a difference of two elements of T_n. When $n \geq 16$, we have $\lfloor\sqrt{n}\rfloor^2 > (\sqrt{n} - 1)^2 \geq n/2 + 1$, so every integer from 0 to $\lfloor n/2 \rfloor$ is a difference of elements of T_n. But then their opposites are also differences, and thus all the integers m satisfying $-n/2 < m \leq n/2$ are differences of elements in T_n. Since every k in $\mathbf{Z}_n$ is equal (mod n) to such an integer m, T_n is a difference set. The set T_n has $2\lfloor\sqrt{n}\rfloor - 1 < 2\sqrt{n}$ elements, and so we have $f(n) < 2\sqrt{n}$, for $n \geq 16$. This implies

$$\frac{\ln f(n)}{\ln n} < \frac{1}{2} + \frac{\ln 2}{\ln n}.$$

We can now use the squeeze principle to conclude that

$$\lim_{n\to\infty} \frac{\ln f(n)}{\ln n} = \frac{1}{2}.$$

Problem 123

a. If a rational function (a quotient of two real polynomials) takes on rational values for infinitely many rational numbers, prove that it may be expressed as the quotient of two polynomials with rational coefficients.

b. If a rational function takes on integral values for infinitely many integers, prove that it must be a polynomial with rational coefficients.

Solution 1.   a.   Let $f(x)/g(x)$ be the rational function. We will prove the result by induction on the sum of the degrees, $\deg f + \deg g$. If $f(x) = 0$ or $\deg f + \deg g = 0$, then the function is constant and the result is immediate. Now suppose it is true for $\deg f + \deg g \leq k$, and let $\deg f + \deg g = k + 1$. We will soon find it convenient to have $\deg f \geq \deg g$; if necessary, this can be arranged by replacing $f(x)/g(x)$ by its reciprocal $g(x)/f(x)$, which will not affect the rationality of any values except for the finitely many roots of $f(x)$. Assuming $\deg f \geq \deg g$, let r_1 be any rational number for which the value $f(r_1)/g(r_1)$ is also rational, say $f(r_1)/g(r_1) = r_2$. Then the polynomial $f(x) - r_2 g(x)$ has r_1 as a root, so there is a polynomial $h(x)$ for which $f(x) - r_2 g(x) = (x - r_1)h(x)$ and hence

$$\frac{f(x)}{g(x)} - r_2 = (x - r_1)\frac{h(x)}{g(x)}.$$

We see that $h(x)/g(x)$ will be rational whenever x and $f(x)/g(x)$ are both rational, except perhaps for $x = r_1$. Therefore, $h(x)/g(x)$ takes on rational values for infinitely many rational x, so we can apply the induction hypothesis to $h(x)/g(x)$ provided $\deg h + \deg g \le k$. But $f(x) - r_2 g(x) = (x - r_1)h(x)$ implies that $\deg h = \deg(f - r_2 g) - 1$, and since $\deg f \ge \deg g$, $\deg(f - r_2 g) \le \deg f$, so that $\deg h \le \deg f - 1$. Therefore, $\deg h + \deg g \le \deg f + \deg g - 1 = k$, the induction hypothesis applies, and $h(x)/g(x)$ is a quotient of polynomials with rational coefficients. But then so is

$$\frac{f(x)}{g(x)} = (x - r_1)\frac{h(x)}{g(x)} + r_2,$$

and we are done.

b.   By (a), we can assume that the rational function is in the form $f(x)/g(x)$, where $f(x)$ and $g(x)$ are polynomials with rational coefficients. Using the division algorithm, we can write

$$\frac{f(x)}{g(x)} = q(x) + \frac{r(x)}{g(x)},$$

where $q(x), r(x)$ are polynomials with rational coefficients and either $r(x) = 0$ or $\deg r < \deg g$. Let M be the least common multiple of the denominators of the coefficients of $q(x)$. (If $q(x) = 0$, set $M = 1$.) Then for any integer n, $q(n)$ is either an integer or at least $1/M$ removed from the closest integer. On the other hand,

$$\lim_{|x| \to \infty} \frac{r(x)}{g(x)} = 0.$$

Thus, for integers n with $|n|$ large enough, $|r(n)/g(n)| < 1/M$, so that

$$\frac{f(n)}{g(n)} = q(n) + \frac{r(n)}{g(n)}$$

can only be an integer if $q(n)$ is an integer and $r(n) = 0$. However, this implies that the polynomial $r(x)$ has infinitely many roots, so $r(x)$ is identically zero, $f(x)/g(x) = q(x)$, and we are done.

Solution 2.   a.   Suppose $f(x)/g(x)$ is a nonzero rational function taking on rational values for the infinitely many rational numbers $x_1, x_2, x_3, \ldots$. We can take $f(x)$ and $g(x)$ to be relatively prime. Write $f(x) = a_m x^m + \cdots + a_0$, $g(x) = b_n x^n + \cdots + b_0$, and $y_i = f(x_i)/g(x_i)$, $i = 1, 2, 3, \ldots$. Then the coefficients $a_m, \ldots, a_0, b_n, \ldots, b_0$ satisfy the infinite, homogeneous linear system

$$\sum_{j=0}^{m} x_i^j a_j - \sum_{j=0}^{n} y_i x_i^j b_j = 0, \qquad i = 1, 2, 3, \ldots. \qquad (*)$$

In fact, we claim that the solution $(a_m, \ldots, a_0, b_n, \ldots, b_0)$ of $(*)$ is unique up to scalar multiples. If $(a_m^*, \ldots, a_0^*, b_n^*, \ldots, b_0^*)$ is another solution, then for $f^*(x) = a_m^* x^m + \cdots + a_0^*$, $g^*(x) = b_n^* x^n + \cdots + b_0^*$ we have

$$\frac{f^*(x_i)}{g^*(x_i)} = \frac{f(x_i)}{g(x_i)},$$

so

$$f^*(x_i)g(x_i) - f(x_i)g^*(x_i) = 0$$

for all i. This implies that $f^*(x)g(x) - f(x)g^*(x)$, having infinitely many roots, is identically zero, so $f^*(x)/g^*(x) = f(x)/g(x)$. Since $f(x)$ and $g(x)$ are relatively prime and the degrees of $f^*(x)$ and $g^*(x)$ are at most those of $f(x)$ and $g(x)$, respectively, it follows that $f^*(x) = C f(x)$ and $g^*(x) = C g(x)$ for some constant C. Then the solution $(a_m^*, \ldots, a_0^*, b_n^*, \ldots, b_0^*)$ is C times $(a_m, \ldots, a_0, b_n, \ldots, b_0)$. We now know that the null space (set of solutions) of $(*)$ is a one-dimensional subspace of $\mathbf{R}^{m+n+2}$. Therefore, there exists a positive integer k for which the finite system

$$\sum_{j=0}^{m} x_i^j a_j - \sum_{j=0}^{n} y_i x_i^j b_j = 0, \qquad i = 1, 2, 3, \ldots, k, \qquad (**)$$

has that same one-dimensional null space. Because the x_i and y_i are rational, the Gaussian elimination method applied to $(**)$ then shows that the null space consists of the multiples of a single *rational* vector. Since $(a_m, \ldots, a_0, b_n, \ldots, b_0)$ is in the null space, it follows that for some nonzero number c, $c f(x)$ and $c g(x)$ have rational coefficients. If we then write

$$\frac{f(x)}{g(x)} = \frac{c f(x)}{c g(x)},$$

we are done.

b.   Let $f(x)/g(x)$ take on integer values for infinitely many integers. By (a), we can assume that $f(x)$ and $g(x)$ have rational coefficients; after multiplying both by a suitable integer, we can even assume they have integer coefficients. We can also take $f(x)$ and $g(x)$ to be relatively prime. Using the Euclidean algorithm and clearing denominators then yields polynomials $p_1(x)$ and $p_2(x)$ with integral coefficients and a positive integer D such that

$$p_1(x) f(x) + p_2(x) g(x) = D.$$

We then have

$$p_1(x) \frac{f(x)}{g(x)} + p_2(x) = \frac{D}{g(x)},$$

so by the given, there are infinitely many integers n for which $D/g(n)$ is an integer.

But if $g(x)$ is not constant, we can have $|g(n)| \leq D$ for at most finitely many integers n, a contradiction. Thus $g(x)$ must be a rational number, $f(x)/g(x)$ is a rational polynomial, and we are done.

Comment. It is well known that even if a polynomial takes on integral values for *all* integers, it may not have integral coefficients. In fact, it can be any integral linear combination of the "binomial polynomials"

$$\binom{x}{n} = \frac{x(x-1)\cdots(x-n+1)}{n!}.$$

However, the polynomial $d!\, p(x)$, where d is the degree of $p(x)$, must have integral coefficients.

Problem 124

Can there be a multiplicative $n \times n$ magic square $(n > 1)$ with entries $1, 2, \ldots, n^2$? That is, does there exist an integer $n > 1$ for which the numbers $1, 2, \ldots, n^2$ can be placed in a square so that the product of all the numbers in any row or column is always the same?

Solution. No. Suppose that there were such a magic square, with the product of the numbers in each row equal to P. Then the product A of the numbers $1, 2, \ldots, n^2$ would be P^n; in particular, A would be an nth power. Now let m be the number of factors of 2 in the prime factorization of A. We will show that m is not divisible by n (provided $n > 1$), so A cannot be an nth power and we will be done.

Note that every second number in the finite sequence $1, 2, \ldots, n^2$ is divisible by 2, which yields $\lfloor n^2/2 \rfloor$ factors of 2 in A, every fourth number is divisible by 4 for another $\lfloor n^2/4 \rfloor$ factors of 2, and so forth. Therefore, we have

$$m = \lfloor n^2/2 \rfloor + \lfloor n^2/4 \rfloor + \cdots + \lfloor n^2/2^k \rfloor,$$

where k is chosen such that $2^k \leq n^2 < 2^{k+1}$. In particular, for $n = 2, 3, 4, 5, 6$ we find $m = 2 + 1 = 3$, $m = 4 + 2 + 1 = 7$, $m = 8 + 4 + 2 + 1 = 15$, $m = 12 + 6 + 3 + 1 = 22$, $m = 18 + 9 + 4 + 2 + 1 = 34$, respectively, and in each case m is not divisible by n. We now give a general proof that for $n > 6$, we have $n(n-1) < m < n^2$, so that m is between successive multiples of n and hence cannot be divisible by n.

First of all, we see from the expression for m above that

$$m \leq n^2/2 + n^2/4 + \cdots + n^2/2^k = n^2 \left(1/2 + 1/4 + \cdots + 1/2^k\right)$$
$$< n^2 \left(1/2 + 1/4 + \cdots + 1/2^k + \cdots\right)$$
$$= n^2,$$

so we do indeed have $m < n^2$.

On the other hand, $\lfloor x \rfloor > x - 1$ for all x, so we have

$$m > (n^2/2 - 1) + (n^2/4 - 1) + \cdots + (n^2/2^k - 1)$$
$$= n^2(1/2 + 1/4 + \cdots + 1/2^k) - k$$
$$= n^2(1 - 1/2^k) - k.$$

To show that this is greater than $n(n-1)$, it is enough to show $n^2/2^k + k < n$, and, since $n^2/2^k < 2$ (by the definition of k) and k is an integer, it is sufficient to show $k < n - 1$.

From $2^k \leq n^2$ we have $k \leq 2 \log_2 n$. It is easy to see, by taking the derivative, that the difference $2 \log_2 n - (n - 1)$ is decreasing for $n \geq 3$. Since $2 \log_2 n < n - 1$ for $n = 7$, we will have $k \leq 2 \log_2 n < n - 1$ for any $n > 6$, and we are done.

Comment. One can see much more quickly that A is not an nth power by using Bertrand's postulate, the theorem that for any integer $N > 1$, there is a prime between N and $2N$.

Problem 125

Note that if the edges of a regular octahedron have length 1, then the distance between any two of its vertices is either 1 or $\sqrt{2}$. Are there other possible configurations of six points in $\mathbf{R}^3$ for which the distance between any two of the points is either 1 or $\sqrt{2}$? If so, find them.

Solution. There is one other such configuration, consisting of the vertices of a rectangular prism whose triangular faces are equilateral triangles of side 1 and whose rectangular faces are squares. One way of positioning this prism in $\mathbf{R}^3$ is shown in Figure 55.

Let X be any configuration of six points in $\mathbf{R}^3$ such that the distance between any two points of X is either 1 or $\sqrt{2}$. Our proof that the points of X must form the vertices either of a prism as described above or of a regular octa-

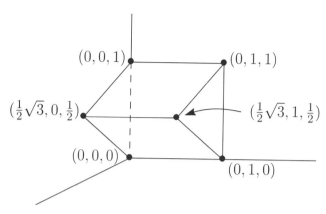

FIGURE 55

hedron will consist of three parts, as follows. In Part 1 we show by contradiction that there are two points of X whose distance is 1. Part 2 is the heart of the proof; here we show that any two points of X whose distance is 1 are vertices of a square whose other two vertices are also in X. Finally, in Part 3 we combine the results of the earlier parts and complete the proof.

Suppose that no two points of X have distance 1, so all distances are $\sqrt{2}$. Then any three points of X form an equilateral triangle. Given such a triangle, there are only two points in $\mathbf{R}^3$ whose distance to each of the vertices is $\sqrt{2}$. But there are three points of X besides the vertices of the triangle, a contradiction. So there are two points of X whose distance is 1, and Part 1 is done.

Now suppose we have any two points of X whose distance is 1. We can position these points at $(0,0,0)$ and $(0,0,1)$. Then in order to have distances from each of these that are either 1 or $\sqrt{2}$, each of the other four points of X must lie on one of the following circles.

C_1: $z = 0$, $x^2 + y^2 = 1$ (distance 1 from $(0,0,0)$, $\sqrt{2}$ from $(0,0,1)$),

C_2: $z = \frac{1}{2}$, $x^2 + y^2 = \frac{3}{4}$ (distance 1 from $(0,0,0)$, 1 from $(0,0,1)$),

C_3: $z = \frac{1}{2}$, $x^2 + y^2 = \frac{7}{4}$ (distance $\sqrt{2}$ from $(0,0,0)$, $\sqrt{2}$ from $(0,0,1)$),

C_4: $z = 1$, $x^2 + y^2 = 1$ (distance $\sqrt{2}$ from $(0,0,0)$, 1 from $(0,0,1)$).

In order to show that $(0,0,0)$ and $(0,0,1)$ are vertices of a square with two other points of X, it will be helpful to consider the projections of the four other points of X onto the xy-plane. In fact, we will show that two of these projections must coincide. This can only happen if one of the points is on C_1 and the other is directly above it on C_4, and we then have our square.

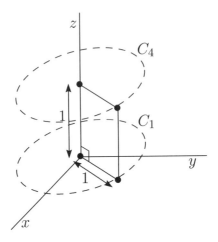

FIGURE 56

Let O be the origin, and let P and Q, with $P \neq Q$, be any two of the projections described above. We can use the law of cosines in $\triangle OPQ$ to get bounds on the angle between the rays OP, OQ from the origin to P, Q respectively. Specifically, if $\theta = \angle POQ$, then we have $PQ^2 = OP^2 + OQ^2 - 2 \cdot OP \cdot OQ \cos \theta$, so

$$\cos \theta = \frac{OP^2 + OQ^2 - PQ^2}{2 \cdot OP \cdot OQ}.$$

Now PQ is at most the distance between the two points of X whose projections are P and Q, so $PQ \leq \sqrt{2}$. On the other hand, those points of X each lie on one of the circles C_1, C_2, C_3, C_4, and so OP and OQ can only have the values $1, \frac{1}{2}\sqrt{3}, \frac{1}{2}\sqrt{7}$. Therefore,

$$OP^2 + OQ^2 - PQ^2 \geq \frac{3}{4} + \frac{3}{4} - 2 = -\frac{1}{2}$$

and thus

$$\cos \theta \geq \frac{-1}{4 \cdot OP \cdot OQ} \geq \frac{-1}{4 \cdot \frac{1}{2}\sqrt{3} \cdot \frac{1}{2}\sqrt{3}} = -\frac{1}{3}.$$

In particular, $\theta < 120°$.

To get a bound in the other direction, we distinguish a number of cases, depending on the location of the points of X whose projections are P and Q. If both these points of X have the same z-coordinate, then PQ equals their distance, so $PQ \geq 1$ and

$$\cos\theta \leq \frac{OP^2 + OQ^2 - 1}{2 \cdot OP \cdot OQ}.$$

We can then have $OP = OQ = 1$, $OP = OQ = \frac{1}{2}\sqrt{3}$, $OP = OQ = \frac{1}{2}\sqrt{7}$, or OP, OQ equal to $\frac{1}{2}\sqrt{3}$, $\frac{1}{2}\sqrt{7}$ in some order. These cases lead to the respective estimates $\cos\theta \leq \frac{1}{2}$, $\cos\theta \leq \frac{1}{3}$, $\cos\theta \leq \frac{5}{7}$, $\cos\theta \leq \frac{3}{\sqrt{21}}$. Therefore, if both the points of X have the same z-coordinate, then $\cos\theta \leq \frac{5}{7}$.

If the points of X whose projections are P and Q have different z-coordinates, then either one of them is on C_1 and the other is on C_4, or their z-coordinates differ by $\frac{1}{2}$. In the former case, since $P \neq Q$, the distance between the points of X must be $\sqrt{2}$ rather than 1, and by the Pythagorean theorem we must then have $PQ = 1$. Then it follows, as above, that $\cos\theta \leq \frac{5}{7}$. In the latter case, since the points of X are at least 1 apart, by the Pythagorean theorem we have $PQ \geq \frac{1}{2}\sqrt{3}$ and thus

$$\cos\theta \leq \frac{OP^2 + OQ^2 - \frac{3}{4}}{2 \cdot OP \cdot OQ}.$$

Meanwhile, OP and OQ are equal to 1, $\frac{1}{2}\sqrt{3}$ or to 1, $\frac{1}{2}\sqrt{7}$ in some order, which yields the estimates $\cos\theta \leq \frac{1}{\sqrt{3}}$, $\cos\theta \leq \frac{2}{\sqrt{7}}$ respectively. Since $\frac{2}{\sqrt{7}} > \frac{5}{7} > \frac{1}{\sqrt{3}}$, we can conclude that $\cos\theta \leq \frac{2}{\sqrt{7}}$ in all cases.

Now consider the four rays from the origin to the projections of the points of X. We have just seen that if the projections of two points of X are distinct, the corresponding rays make an angle θ for which $-\frac{1}{3} \leq \cos\theta \leq \frac{2}{\sqrt{7}}$. Suppose that as we go around the origin, the angles between successive rays are α, β, γ, δ, as shown in Figure 57.

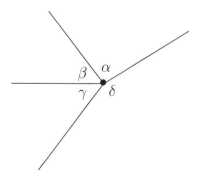

FIGURE 57

We can assume that $\alpha + \beta \leq 180°$ and $\beta + \gamma \leq 180°$ without loss of generality. But then, by the above, we actually have $\alpha + \beta < 120°$ and $\beta + \gamma < 120°$, so that

$$\delta = 360° - (\alpha + \beta + \gamma) \geq 360° - (\alpha + \beta) - (\beta + \gamma) > 120°.$$

Therefore, the smallest angle between the rays that form δ must be $\alpha + \beta + \gamma$ rather than δ, and we have $\alpha + \beta + \gamma < 120°$. On the other hand, if all four rays are distinct, then $\alpha, \beta, \gamma \geq \cos^{-1} \left(\frac{2}{\sqrt{7}} \right)$, and using the formula

$$\cos 3\theta = 4 \cos^3 \theta - 3 \cos \theta,$$

we find

$$\cos(\alpha + \beta + \gamma) \leq \cos \left(3 \cos^{-1} \left(\frac{2}{\sqrt{7}} \right) \right) = \frac{-10}{7\sqrt{7}} < -\frac{1}{2},$$

a contradiction. We have now shown that two of the projections must coincide, so $(0, 0, 0)$ and $(0, 0, 1)$ are vertices of a square with two other points of X, and Part 2 is done.

By Parts 1 and 2, we know that there are four points of X which form the vertices of a square of side 1. Let P_1 and P_2 be the remaining two points of X. P_1 and P_2 cannot both have distance $\sqrt{2}$ to each vertex of the square, since the only two points in $\mathbf{R}^3$ with this property lie at distance $\sqrt{6}$ to each other, one on either side of the plane that the square is in. So we can assume that P_1 has distance 1 to one of the vertices of the square, say to P_3. But then by Part 2 of our proof, there is another square whose vertices are points of X and include P_1 and P_3. Since two distinct squares cannot have three vertices in common, the new square can have at most two of the vertices of the old square, and since X has only six points in all, it follows that *both* P_1 and P_2 must be vertices of the new square.

First suppose that P_1 and P_2 are adjacent vertices of the new square. Then the two squares have a side in common, so they are "hinged" as shown in Figure 58. By considering the distances from P_1 to P_4 and P_5, we see that $P_1 P_3 P_4$ must be an equilateral triangle, and so we have the prism.

On the other hand, if P_1 and P_2 are opposite vertices of the new square, then the two squares have a diagonal in common (see Figure 59). Since the distances from P_2 and from P_4 to the midpoint M of this diagonal are both $\frac{1}{2}\sqrt{2}$, the distance from P_2 to P_4 cannot be $\sqrt{2}$, so it must be 1. Then $P_2 M$ and $P_4 M$ are perpendicular, and we have the regular octahedron.

Finally, the prism and the regular octahedron yield distinct configurations, because for the prism each vertex has distance $\sqrt{2}$ from two of the other ver-

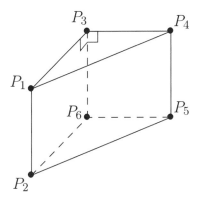

FIGURE 58

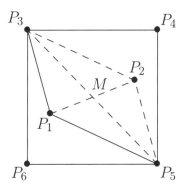

FIGURE 59

tices, while for the octahedron each vertex has distance $\sqrt{2}$ from only one other vertex.

Problem 126

Let a and b be positive real numbers, and define a sequence (x_n) by

$$x_0 = a, \quad x_1 = b, \quad x_{n+1} = \frac{1}{2}\left(\frac{1}{x_n} + x_{n-1}\right).$$

a. For what values of a and b will this sequence be periodic?

b. Show that given a, there exists a unique b for which the sequence converges.

Solution. a. The sequence is periodic if and only if $ab = 1$. To show this, we can rewrite the given recurrence relation as

$$x_n x_{n+1} = \frac{1}{2}\left(1 + x_{n-1} x_n\right).$$

If we then put $y_n = x_n x_{n+1}$, we see that the sequence $(y_n)_{n \geq 0}$ satisfies the recurrence relation

$$y_{n+1} = \frac{1}{2}\left(1 + y_n\right).$$

Also, if (x_n) is periodic, then so is (y_n). On the other hand, if $y_n < 1$, then $y_n < y_{n+1} < 1$, and if $y_n > 1$, then $y_n > y_{n+1} > 1$, so (y_n) is not periodic if $y_n \neq 1$. If $y_n = 1$, then $y_{n+1} = 1$. Therefore, (y_n) is periodic if and only if $y_0 = 1$, that is, $ab = 1$. In this case, the original sequence (x_n) has the form $a, 1/a, a, 1/a, \ldots$, so it is periodic as well.

  b. First note that if (x_n) converges to L, then the relation

$$x_n x_{n+1} = \frac{1}{2}(1 + x_{n-1} x_n)$$

implies $L^2 = \frac{1}{2}\left(1 + L^2\right)$, and so $L = 1$ (since $L \geq 0$).

  To find out when this happens, we use the sequence (y_n) from (a) to get explicit expressions for the x_n. Rewriting the recurrence relation for (y_n) as

$$(y_{n+1} - 1) = \frac{1}{2}\left(y_n - 1\right),$$

we find by induction that

$$y_n - 1 = \frac{1}{2^n}\left(y_0 - 1\right),$$

so

$$y_n = 1 + \frac{y_0 - 1}{2^n} = 1 + \frac{ab - 1}{2^n}.$$

Now

$$\frac{x_{n+2}}{x_n} = \frac{y_{n+1}}{y_n} = \frac{1 + \dfrac{ab - 1}{2^{n+1}}}{1 + \dfrac{ab - 1}{2^n}}.$$

Therefore, by induction,

$$x_{2n} = \left(\prod_{j=0}^{n-1} \frac{1 + \dfrac{ab - 1}{2^{2j+1}}}{1 + \dfrac{ab - 1}{2^{2j}}}\right) a$$

and

$$x_{2n+1} = \left(\prod_{j=1}^{n} \frac{1 + \dfrac{ab - 1}{2^{2j}}}{1 + \dfrac{ab - 1}{2^{2j-1}}} \right) b \,.$$

If (x_n) converges, then

$$\lim_{n \to \infty} x_{2n} = \left(\prod_{j=0}^{\infty} \frac{1 + \dfrac{ab - 1}{2^{2j+1}}}{1 + \dfrac{ab - 1}{2^{2j}}} \right) a = L = 1,$$

so

$$\prod_{j=0}^{\infty} \frac{1 + \dfrac{ab - 1}{2^{2j+1}}}{1 + \dfrac{ab - 1}{2^{2j}}} = \frac{1}{a} \,.$$

Conversely, if

$$\prod_{j=0}^{\infty} \frac{1 + \dfrac{ab - 1}{2^{2j+1}}}{1 + \dfrac{ab - 1}{2^{2j}}} = \frac{1}{a} \,,$$

then we not only have

$$\lim_{n \to \infty} x_{2n} = 1 \,,$$

but also

$$\lim_{n \to \infty} x_{2n+1} = \left(\frac{1 + \dfrac{ab - 1}{2^2}}{1 + \dfrac{ab - 1}{2}} \cdot \frac{1 + \dfrac{ab - 1}{2^4}}{1 + \dfrac{ab - 1}{2^3}} \cdot \cdots \right) b$$

$$= \frac{1}{1 + (ab - 1)} \left(\frac{1 + \dfrac{ab - 1}{2}}{1 + ab - 1} \cdot \frac{1 + \dfrac{ab - 1}{2^3}}{1 + \dfrac{ab - 1}{2^2}} \cdot \cdots \right)^{-1} b$$

$$= \frac{1}{ab} \left(\prod_{j=0}^{\infty} \frac{1 + \dfrac{ab - 1}{2^{2j+1}}}{1 + \dfrac{ab - 1}{2^{2j}}} \right)^{-1} b = 1,$$

and hence (x_n) converges.

For fixed a, the value of the convergent infinite product

$$\prod_{j=0}^{\infty} \frac{1 + \dfrac{ab-1}{2^{2j+1}}}{1 + \dfrac{ab-1}{2^{2j}}}$$

is a continuous function of b for $b > 0$. As b increases from 0 to ∞, the individual factors

$$\frac{1 + \dfrac{ab-1}{2^{2j+1}}}{1 + \dfrac{ab-1}{2^{2j}}} = \frac{2^{2j+1} - 1 + ab}{2^{2j+1} - 2 + 2ab}$$

of the product decrease from

$$\begin{cases} (2^{2j+1} - 1)/(2^{2j+1} - 2), & j > 0 \\ \\ \infty, & j = 0 \end{cases}$$

to $1/2$. Thus the product is a decreasing function of b, whose limit as $b \to \infty$ is 0, and whose limit as $b \to 0^+$ is ∞. Therefore, given a, there is a unique b for which the product equals $1/a$, and we are done.

Comment. One systematic way to find the general formula for y_n is to introduce a generating function, as follows. Let

$$g(z) = \sum_{n=0}^{\infty} y_n z^n.$$

Then from the recurrence relation $y_{n+1} = \frac{1}{2}(1 + y_n)$, we find that

$$g(z) = y_0 + z \sum_{n=0}^{\infty} \frac{1}{2}(1 + y_n) z^n$$

$$= y_0 + \frac{z/2}{1-z} + \frac{z}{2} g(z).$$

Solving for $g(z)$ yields

$$g(z) = \frac{y_0}{1 - z/2} + \frac{z/2}{(1 - z/2)(1 - z)}$$

$$= \frac{y_0 - 1}{1 - z/2} + \frac{1}{1 - z}$$

$$= \sum_{n=0}^{\infty} \left[\frac{1}{2^n}(y_0 - 1) + 1 \right] z^n,$$

and we can now read off that

$$y_n = \frac{1}{2^n}(y_0 - 1) + 1.$$

Problem 127

Consider the equation $x^2 + \cos^2 x = \alpha \cos x$, where α is some positive real number.

a. For what value or values of α does the equation have a unique solution?

b. For how many values of α does the equation have precisely four solutions?

Idea. Solutions of the equation correspond to intersection points of the circle $x^2 + y^2 = \alpha y$ and the graph of $y = \cos x$.

Solution. a. The equation has a unique solution if and only if $\alpha = 1$.

To see this, note that if x is a solution, so is $-x$; therefore, if the solution is unique, it must be $x = 0$. If $x = 0$ is a solution to $x^2 + \cos^2 x = \alpha \cos x$, then $\alpha = 1$.

It remains to show that when $\alpha = 1$, $x = 0$ is the *only* solution. If $x^2 + \cos^2 x = \cos x$, then $x^2 = \cos x - \cos^2 x \leq 1 - \cos^2 x = \sin^2 x$, so $|x| \leq |\sin x|$, and since this is true only for $x = 0$, we are done.

b. There is exactly one $\alpha > 0$ for which the equation has precisely four solutions.

If we put $y = \cos x$, the given equation becomes $x^2 + y^2 = \alpha y$ or, completing the square, $x^2 + (y - \frac{1}{2}\alpha)^2 = \frac{1}{4}\alpha^2$. Since $\alpha > 0$, this represents a circle with center $(0, \frac{1}{2}\alpha)$ and radius $\frac{1}{2}\alpha$. Thus the solutions of our equation $x^2 + \cos^2 x = \alpha \cos x$ correspond to intersection points of the graph of $y = \cos x$ with the circle $x^2 + (y - \frac{1}{2}\alpha)^2 = \frac{1}{4}\alpha^2$. Figure 60 shows such circles for various values of α, along with the graph of $y = \cos x$. Our problem can be reformulated as: "How many of these circles intersect the graph in precisely four points?"

In part (a) we saw that the circle for $\alpha = 1$ lies underneath the graph of $y = \cos x$ except for $x = 0$, where it is tangent to the graph. Therefore, for any $\alpha < 1$ the circle $x^2 + (y - \frac{1}{2}\alpha)^2 = \frac{1}{4}\alpha^2$, which lies inside the circle for $\alpha = 1$, does not intersect $y = \cos x$ at all. Hence it is enough to consider $\alpha > 1$.

Figure 60 suggests that for any $\alpha > 1$, there are two intersection points with $-\pi/2 < x < \pi/2$. In fact, since the point $(0, 1)$ on the graph is inside the circle while the points $(-\pi/2, 0)$ and $(\pi/2, 0)$ are outside, there must be at

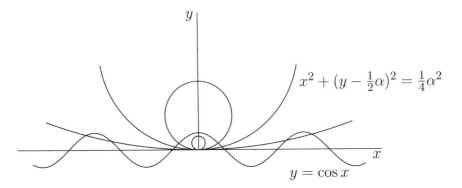

$$x^2 + (y - \tfrac{1}{2}\alpha)^2 = \tfrac{1}{4}\alpha^2$$

$$y = \cos x$$

FIGURE 60

least two intersection points. On the other hand, if there were more than two, the function $f(x) = x^2 + \cos^2 x - \alpha \cos x$ would have at least three zeros with $-\pi/2 < x < \pi/2$. Applying Rolle's theorem to $f(x)$ and then to $f'(x)$, we see that $f''(x) = 0$ would have a solution in this same interval. But

$$f''(x) = 2 - 2\cos 2x + \alpha \cos x \ge \alpha \cos x > 0$$

for $-\pi/2 < x < \pi/2$, a contradiction. Thus we have shown that if $\alpha > 1$, there are precisely two intersection points with $-\pi/2 < x < \pi/2$.

If, for some α, there are also two intersection points with $3\pi/2 < x < 5\pi/2$, the next interval where $\cos x > 0$, then by symmetry there will be two more for $-5\pi/2 < x < -3\pi/2$, making at least six intersection points altogether. On the other hand, suppose there are *no* intersection points for $3\pi/2 < x < 5\pi/2$. Then the circle must pass above the point $(2\pi, 1)$ (if, indeed, the circle extends that far to the right) and thus there will be no intersection points for $|x| > 2\pi$ either. This leaves us with only two intersection points in all.

Thus the only possible way to get precisely four intersection points is to have a *unique* intersection point with $3\pi/2 < x < 5\pi/2$. We now show that this happens for exactly one value of α.

Note that for each point on the graph of $y = \cos x$ with $3\pi/2 < x < 5\pi/2$, there is exactly one of the circles $x^2 + (y - \tfrac{1}{2}\alpha)^2 = \tfrac{1}{4}\alpha^2$ which passes through that point; specifically, the one with

$$\alpha = \frac{x^2 + y^2}{y} = \frac{x^2 + \cos^2 x}{\cos x}.$$

If we put

$$F(x) = \frac{x^2 + \cos^2 x}{\cos x},$$

then we want to show that there is exactly one value of α for which the equation $F(x) = \alpha$ has a unique solution with $3\pi/2 < x < 5\pi/2$.

Now we have $F(x) > 0$ for $3\pi/2 < x < 5\pi/2$, and $F(x) \to \infty$ as $x \to (3\pi/2)^+$ and as $x \to (5\pi/2)^-$. Therefore, $F(x)$ has a minimum value α_0 on the interval $3\pi/2 < x < 5\pi/2$; by the Intermediate Value Theorem, $F(x)$ takes on all values greater than α_0 at least *twice* on that interval.

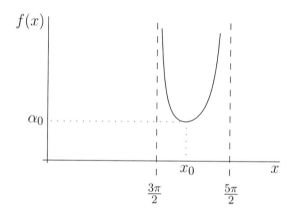

FIGURE 61

Hence the only value of α for which $F(x) = \alpha$ can actually have a unique solution in the interval is $\alpha = \alpha_0$. To show that this is indeed the case, it is enough to show that $F'(x)$ is zero only once on the interval, since then the minimum value can occur only once.

Now

$$F'(x) = \frac{(2x - 2 \sin x \cos x) \cos x - (x^2 + \cos^2 x)(-\sin x)}{\cos^2 x}$$

can only be zero when its numerator is, that is, when

$$x^2 \sin x + 2x \cos x - \sin x \cos^2 x = 0.$$

On the other hand, the derivative of $g(x) = x^2 \sin x + 2x \cos x - \sin x \cos^2 x$ is

$$g'(x) = (x^2 + 2 + 2 \sin^2 x - \cos^2 x) \cos x = (x^2 + 4 - 3 \cos^2 x) \cos x,$$

which is positive on our interval. So $g(x)$ is increasing there and can only be zero once, and it follows that there is a unique x_0 in the interval with $F(x_0) = \alpha_0$.

Finally, since the circle $x^2 + (y - \frac{1}{2}\alpha_0)^2 = \frac{1}{4}\alpha_0^2$ lies above the graph of $y = \cos x$ at both end points $3\pi/2$, $5\pi/2$ of the interval (provided the circle extends that far to the right), it must be above the graph everywhere in the interval except at the point of tangency $(x_0, \cos x_0)$. But then by the periodicity of cosine, the circle will certainly lie above the graph on all "future" intervals $7\pi/2 < x < 9\pi/2, \ldots$ where $\cos x > 0$. Thus for $\alpha = \alpha_0$ there are precisely four intersection points, for any other value of α there is a different number of intersection points, and we are done.

Problem 128

Fast Eddie needs to double his money; he can only do so by playing a certain win-lose game, in which the probability of winning is p. However, he can play this game as many or as few times as he wishes, and in a particular game he can bet any desired fraction of his bankroll. The game pays even money (the odds are one-to-one). Assuming he follows an optimal strategy if one is available, what is the probability, as a function of p, that Fast Eddie will succeed in doubling his money?

Solution. The probability is 1 for $p > 1/2$ and p for $p \le 1/2$.

Suppose that $p > 1/2$; we will show a way for Fast Eddie to double his money with probability 1. To begin, suppose that at some point his bankroll has reached x, and that he proceeds to bet $x/2^n$, $2x/2^n$, $4x/2^n$, $\ldots$ until he wins once or until he has lost n of these bets. Then if he wins once, his bankroll will become $x(1 + \frac{1}{2^n})$, and this will occur with probability $1 - (1 - p)^n$. If he does win once and he continues with a similar round of up to n bets, but with x replaced by the new bankroll $x(1 + \frac{1}{2^n})$, and if Eddie proceeds in this way for a total of 2^n rounds (winning once in each round), his bankroll will reach

$$x\left(1 + \frac{1}{2^n}\right)^{2^n} > 2x.$$

In particular, if x was Eddie's original bankroll, he will have doubled it after the 2^n rounds. Meanwhile, the probability that he will indeed win once in each round is

$$\left(1 - (1 - p)^n\right)^{2^n}.$$

Note that since $p > 1/2$, this approaches 1 as $n \to \infty$. (This can be seen, for instance, by taking the logarithm, rewriting the resulting indeterminate form of type $\infty \cdot 0$ as a quotient, and using l'Hôpital's rule.) So by choosing a large n, we can find a strategy for which Eddie's probability of doubling his bankroll (after at most $N = n\,2^n$ bets) is as close to 1 as desired. However, since n has to be chosen in advance, we cannot actually achieve probability 1 this way, so we do not yet have an optimal strategy.

To get an optimal strategy, we use a refinement of the same idea. If we replace "2^n rounds" in the discussion above by "$2^n M$ rounds," where M is a positive integer, Eddie's bankroll will grow from x to an amount $> 2^M x$ with probability $(1 - (1 - p)^n)^{2^n M}$, which still approaches 1 as $n \to \infty$. So even if x is *not* Eddie's original bankroll, he can make his bankroll reach twice its original size (using at most $N = M\,n\,2^n$ bets) with probability as close to 1 as desired.

Now we can see that the following is an optimal strategy for Fast Eddie (still for $p > 1/2$). Let him choose a number N_1 such that the bankroll will double after at most N_1 bets, as described in the first paragraph, with some probability $p_1 > 1/2$, and let him make those bets. If he loses in some round, he will still have at least $1/2^{N_1}$ of his original bankroll. In this case, let him choose a number N_2 such that this amount will grow to twice the original amount after at most N_2 bets, as described in the second paragraph, with some probability $p_2 > 1/2$. Let him then make those bets, and so forth. The total probability of success (in doubling his original bankroll) will then be

$$p_1 + (1 - p_1)p_2 + (1 - p_1)(1 - p_2)p_3 + \cdots.$$

The partial sums of this series are

$$1 - (1 - p_1),\ 1 - (1 - p_1)(1 - p_2),\ 1 - (1 - p_1)(1 - p_2)(1 - p_3),\ldots,$$

so they have limit 1 (since $p_i > 1/2$ for all i), and we are done with the case $p > 1/2$.

For the case $p \le 1/2$, an optimal strategy for Fast Eddie is to bet all his money at once; thus, his probability of success will be p. To show this, suppose that he follows another strategy, for which the probability of success is q. We will show that $q \le p$. In carrying out this strategy, the cumulative amount Eddie must wager is at least his bankroll (otherwise he cannot possibly double his money). Since all his bets have probability $p \le 1/2$, his expected net return is at most $p - (1 - p) = 2p - 1$ times his bankroll. On the other hand, the worst that can happen to Eddie is that he loses his entire bankroll, and so his expected net return is at least $q - (1 - q) = 2q - 1$ times his bankroll. Thus $2q - 1 \le 2p - 1$, so $q \le p$, and we are done.

Comments. This problem may be viewed as a one-dimensional random walk problem, which has unequal probabilities and variable step sizes.

In practice, there will be a minimum bet (a penny, say), and then the probability will *not* be 1, even for $p > 1/2$.

Problem 129

Define a *die* to be a convex polyhedron. For what n is there a fair die with n faces? By fair, we mean that, given any two faces, there exists a symmetry of the polyhedron which takes the first face to the second.

Solution. There is a fair die with n faces if and only if n is an even integer with $n > 2$.

Any convex polyhedron has at least four faces, so we can assume $n \geq 4$. A regular tetrahedron is a fair die with 4 faces. If n is even and $n > 4$, we can get a fair die with n faces by constructing a "generalized regular octahedron," as follows. Begin with a regular $n/2$-gon in a horizontal plane. Place two points, one on either side of the plane, on the vertical line through the center of the $n/2$-gon, at equal distances from the center. Connect these two points to the $n/2$ vertices of the $n/2$-gon to obtain a fair die with n triangular faces.

Now suppose n is odd. We use Euler's formula, $V - E + F = 2$, where V is the number of vertices, E is the number of edges, and F is the number of faces of a simple (no "holes") polyhedron. In our case, $F = n$. Since all faces are congruent, they all have the same number, say s, of sides. Because each edge bounds two faces, $E = sn/2$, and so s is even. Let $v_1 \leq v_2 \leq \cdots \leq v_s$ denote the numbers of edges emanating from the s vertices of a single face. Since a vertex with v_i edges emanating from it is a vertex for v_i different faces, we have

$$V = n \left(\frac{1}{v_1} + \frac{1}{v_2} + \cdots + \frac{1}{v_s} \right).$$

Substituting all this into Euler's formula yields

$$n \left(\frac{1}{v_1} + \frac{1}{v_2} + \cdots + \frac{1}{v_s} - \frac{s}{2} + 1 \right) = 2.$$

Since $v_i \geq 3$ for all i, this implies

$$2 \leq n \left(\frac{s}{3} - \frac{s}{2} + 1 \right) = n \left(1 - \frac{s}{6} \right),$$

and since s is even, it follows that $s = 4$. Euler's formula then becomes

$$n \left(\frac{1}{v_1} + \frac{1}{v_2} + \frac{1}{v_3} + \frac{1}{v_4} - 1 \right) = 2.$$

Since $v_1 \geq 4$ would imply

$$\frac{1}{v_1} + \frac{1}{v_2} + \frac{1}{v_3} + \frac{1}{v_4} \leq 1,$$

we conclude that $v_1 = 3$.

At this point, our goal is to show that, for odd n, the equation

$$\frac{1}{v_2} + \frac{1}{v_3} + \frac{1}{v_4} = \frac{2}{3} + \frac{2}{n}$$

has no solution in positive integers $v_2 \leq v_3 \leq v_4$ with $v_2 \geq 3$.

Observe that

$$\frac{1}{3} + \frac{1}{6} + \frac{1}{6} = \frac{2}{3} < \frac{2}{3} + \frac{2}{n} \quad \text{and} \quad \frac{1}{4} + \frac{1}{5} + \frac{1}{5} = \frac{13}{20} < \frac{2}{3},$$

which leaves us with the following four cases:

i. $v_2 = 3, v_3 = 3$;

ii. $v_2 = 3, v_3 = 4$;

iii. $v_2 = 3, v_3 = 5$; and

iv. $v_2 = 4, v_3 = 4$.

If v_2 and v_3 are both odd, then

$$\frac{1}{v_2} + \frac{1}{v_3} + \frac{1}{v_4} = \frac{v_2 v_3 + v_2 v_4 + v_3 v_4}{v_2 v_3 v_4}$$

has odd numerator, so it cannot equal

$$\frac{2}{3} + \frac{2}{n} = \frac{2(n+3)}{3n};$$

this rules out cases (i) and (iii). Case (ii) reduces to

$$\frac{1}{v_4} = \frac{n+24}{12n}.$$

In this case, then, $n + 24$ is an odd factor of $12n$, hence of $3n$. Since $n + 24$ also divides $3n + 72$, we find that $n + 24$ is an odd factor of 72, which is impossible. Finally, case (iv) reduces to

$$\frac{1}{v_4} = \frac{n+12}{6n},$$

and by a similar argument, this is impossible. Thus there is no fair die with an odd number of faces, and we are done.

Problem 130

Prove that

$$\det \begin{pmatrix} 1 & 4 & 9 & \cdots & n^2 \\ n^2 & 1 & 4 & \cdots & (n-1)^2 \\ \vdots & \vdots & \vdots & \vdots & \vdots \\ 4 & 9 & 16 & \cdots & 1 \end{pmatrix}$$

$$= (-1)^{n-1} \frac{n^{n-2}(n+1)(2n+1)\big((n+2)^n - n^n\big)}{12}.$$

Solution. We begin by sketching a proof of the known formula for the determinant of a circulant matrix:

$$\det \begin{pmatrix} a_1 & a_2 & a_3 & \cdots & a_n \\ a_n & a_1 & a_2 & \cdots & a_{n-1} \\ \vdots & \vdots & \vdots & \vdots & \vdots \\ a_3 & a_4 & a_5 & \cdots & a_2 \\ a_2 & a_3 & a_4 & \cdots & a_1 \end{pmatrix} = (-1)^{n-1} \prod_{j=0}^{n-1} \left(\sum_{k=1}^{n} \zeta^{jk} a_k \right),$$

where $\zeta = e^{2\pi i/n}$.

View the formula above as a polynomial identity in $a_1, \ldots, a_n$. Note that the determinant is zero if $a_1 + a_2 + \cdots + a_n = 0$, since then the sum of all the columns of the matrix is zero. Therefore, the determinant has a factor $a_1 + a_2 + \cdots + a_n$, and this is the factor on the right for $j = 0$. Similarly, if for any j we multiply the first column of the matrix by ζ^j, the second column by ζ^{2j}, and so forth, and add all the columns, we get a multiple of $\sum_{k=1}^{n} \zeta^{jk} a_k$ in each position. Thus if $\sum_{k=1}^{n} \zeta^{jk} a_k = 0$, the original columns are linearly dependent and the determinant is zero. It follows that the determinant has a factor $\sum_{k=1}^{n} \zeta^{jk} a_k$ for every j. Furthermore, no two of these n polynomials are constant multiples of each other. Both sides of the formula above are homogeneous polynomials of total degree n in $a_1, \ldots, a_n$, so there are no nonconstant factors other than the ones we have accounted for already. Finally, we can get the factor $(-1)^{n-1}$ by looking at the case of the identity matrix.

Applying the formula to our special case $a_k = k^2$, we find

$$\det \begin{pmatrix} 1 & 4 & 9 & \cdots & n^2 \\ n^2 & 1 & 4 & \cdots & (n-1)^2 \\ \vdots & \vdots & \vdots & \vdots & \vdots \\ 4 & 9 & 16 & \cdots & 1 \end{pmatrix} = (-1)^{n-1} \prod_{j=0}^{n-1} \left(\sum_{k=1}^{n} \zeta^{jk} k^2 \right).$$

The next step is to derive a "closed" expression for $\sum_{k=1}^{n} k^2 x^k$, after which we will substitute $x = \zeta^j$ for $j = 0, 1, \ldots, n-1$. Specifically, we claim that

$$\sum_{k=1}^{n} k^2 x^k$$
$$= \begin{cases} \dfrac{n^2 x^{n+3} - (2n^2 + 2n - 1)x^{n+2} + (n^2 + 2n + 1)x^{n+1} - x^2 - x}{(x-1)^3} & \text{if } x \neq 1 \\ \dfrac{n(n+1)(2n+1)}{6} & \text{if } x = 1. \end{cases}$$

To obtain this formula for $x \neq 1$, start with the finite geometric series

$$\sum_{k=0}^{n} x^k = \frac{x^{n+1} - 1}{x - 1}.$$

Then differentiate each side and multiply by x, to find

$$\sum_{k=1}^{n} k x^k = \frac{n x^{n+2} - (n+1)x^{n+1} + x}{(x-1)^2}.$$

Differentiating and multiplying by x a second time yields the formula we want. For the well-known special case $x = 1$, we can apply l'Hôpital's rule to the general case. For either case, one could also use induction on n.

We can now make some progress toward the desired formula:

$$\det \begin{pmatrix} 1 & 4 & 9 & \cdots & n^2 \\ n^2 & 1 & 4 & \cdots & (n-1)^2 \\ \vdots & \vdots & \vdots & \vdots & \vdots \\ 4 & 9 & 16 & \cdots & 1 \end{pmatrix} = (-1)^{n-1} \prod_{j=0}^{n-1} \left(\sum_{k=1}^{n} k^2 \zeta^{jk} \right)$$

$$= (-1)^{n-1} \frac{n(n+1)(2n+1)}{6} \prod_{j=1}^{n-1} \frac{n^2 \zeta^{3j} - (2n^2 + 2n)\zeta^{2j} + (n^2 + 2n)\zeta^j}{(\zeta^j - 1)^3}$$

$$= (-1)^{n-1} \frac{n(n+1)(2n+1)}{6} \prod_{j=1}^{n-1} \frac{n^2 \zeta^j (\zeta^j - \frac{n+2}{n})}{(\zeta^j - 1)^2}.$$

To conclude, we use a technique reminiscent of the standard method used to prove the irreducibility of cyclotomic polynomials. To find

$$\prod_{j=1}^{n-1}(\zeta^j - a)$$

(we will later take $a = 0$, $a = 1$, and $a = (n+2)/n$, note that the $\zeta^j - a$ are the roots of the polynomial

$$\frac{(y+a)^n - 1}{(y+a) - 1} = \sum_{k=0}^{n-1}(y+a)^k,$$

so their product is $(-1)^{n-1}$ times the constant term of the polynomial. Thus we have

$$\prod_{j=1}^{n-1}(\zeta^j - a) = (-1)^{n-1}\sum_{k=0}^{n-1}a^k$$

$$= \begin{cases} (-1)^{n-1}\dfrac{a^n - 1}{a - 1} & \text{if } a \neq 1, \\[2mm] (-1)^{n-1}n & \text{if } a = 1. \end{cases}$$

In particular,

$$\prod_{j=1}^{n-1}\zeta^j = (-1)^{n-1}, \qquad \prod_{j=1}^{n-1}(\zeta^j - 1) = (-1)^{n-1}n,$$

and

$$\prod_{j=1}^{n-1}\left(\zeta^j - \frac{n+2}{n}\right) = (-1)^{n-1}\frac{(n+2)^n - n^n}{2n^{n-1}}.$$

Substituting these results into our most recent expression for the determinant and simplifying, we get

$$\det\begin{pmatrix} 1 & 4 & 9 & \cdots & n^2 \\ n^2 & 1 & 4 & \cdots & (n-1)^2 \\ \vdots & \vdots & \vdots & \vdots & \vdots \\ 4 & 9 & 16 & \cdots & 1 \end{pmatrix}$$

$$= (-1)^{n-1}\frac{n^{n-2}(n+1)(2n+1)\big((n+2)^n - n^n\big)}{12},$$

as claimed.

Comment. If we rewrite the determinant as

$$\frac{(-1)^{n-1}n^{2n}\left(1+\frac{1}{n}\right)\left(2+\frac{1}{n}\right)\left[\left(1+\frac{2}{n}\right)^{n}-1\right]}{12},$$

we see that its absolute value is asymptotic to

$$\frac{n^{2n}(e^2-1)}{6}.$$

PREREQUISITES BY PROBLEM NUMBER

1. Basic properties of the integers

2. Elementary algebra

3. Differential calculus

4. Precalculus

5. Geometric series

6. Elementary algebra

7. Precalculus

8. Elementary geometry

9. Elementary algebra

10. Determinants

11. Elementary algebra

12. Integral calculus

13. Elementary probability

14. Limits of sequences

15. Basic properties of the integers; elementary algebra

16. Differential calculus

17. Basic properties of the integers

18. Inclusion-exclusion principle

19. Analytic geometry

20. Vector geometry

21. Precalculus

22. Elementary number theory; geometric series

23. None

24. Elementary geometry

25. None

26. Calculus

27. Basic properties of the integers

28. Elementary algebra

29. Elementary geometry

30. Trigonometry

31. Elementary algebra

32. Differential calculus

33. Elementary algebra

34. Analytic geometry

35. Trigonometry

36. Matrix algebra

37. Elementary geometry

38. Elementary algebra

39. Elementary geometry

40. Calculus

41. Basic properties of the integers; binomial theorem

42. Differential calculus

43. Differential calculus

44. Algebra of complex numbers

45. Elementary geometry; elementary algebra

46. Taylor series

47. Linear algebra

48. Trigonometry

49. Differential calculus

50. Linear algebra

51. Elementary geometry

52. Elementary number theory

53. Basic counting methods

54. Vector arithmetic

55. Elementary geometry

56. Factorization of polynomials

57. Basic counting methods

58. Elementary number theory; mathematical induction

59. Analytic geometry

60. Limits of sequences

61. Vector geometry

62. Precalculus

63. Elementary probability

64. Polar coordinates

65. Elementary number theory

66. Infinite series

67. Vector geometry

68. Calculus

69. Basic counting methods

70. Elementary number theory

71. Limits of sequences

72. Integral calculus

73. Multivariable calculus

74. Analytic geometry; elementary number theory

75. Infinite series

76. Abstract algebra

77. Elementary number theory

78. Limits of sequences; mathematical induction

79. Basic counting methods

80. Linear algebra; calculus

81. Taylor series

82. None

83. Infinite series

84. Indeterminate forms; mathematical induction

85. None

86. Elementary geometry

87. None

88. Indeterminate forms

89. Limits of sequences

90. Infinite series

91. Calculus; definition of limit

92. Taylor series

93. Elementary number theory

94. Basic counting methods

95. Limits of sequences

96. Limits of sequences in the plane

97. Differential calculus

125. Analytic geometry; trigonometry

126. Limits of sequences; infinite products

127. Differential calculus

128. Probability theory; differential calculus

129. Euler's formula for polyhedra

130. Roots of unity; factorization of polynomials; theory of determinants; closed forms for finite sums

PROBLEM NUMBERS BY SUBJECT

ALGEBRA AND TRIGONOMETRY

    Complex numbers   34, 44

    Equations   2, 7, 9, 18, 21, 44, 56, 64, 105, 106; see also CALCULUS

    Functions   4, 62

    Inequalities   5, 11, 27, 28, 31, 33, 121

    Trigonometry   30, 35, 48, 64, 125, 127

CALCULUS

    Equations   7, 16, 49, 105, 127; see also ALGEBRA AND
        TRIGONOMETRY

    Estimation   68, 81, 89, 122; see also Limits

    Integration   12, 26, 40, 68, 72, 81, 91, 99, 102, 108

    Intermediate Value Theorem   40, 51, 91, 107

    Infinite series (convergence)   75, 83, 90, 118

    Infinite series (evaluation)   46, 66, 115

    Infinite series (Taylor series)   46, 81, 92, 104

    Limits   60, 84, 88, 89, 91, 96, 113, 122

    Limits (iteration, recursion)   14, 71, 78, 95, 101, 107, 119, 126

    Maximum-minimum problems   3, 48

    Multivariable calculus   73, 99

    Probability   99, 128; see also DISCRETE MATHEMATICS

    Tangent lines   32, 42, 43, 97

INDEX

n : occurs in the statement or the solution(s) of problem n
n^C : occurs only in the comment of problem n
n^I : occurs only in the idea of problem n
n^m : occurs in solution m of problem n